ANATOMY
ILLUSTRATED

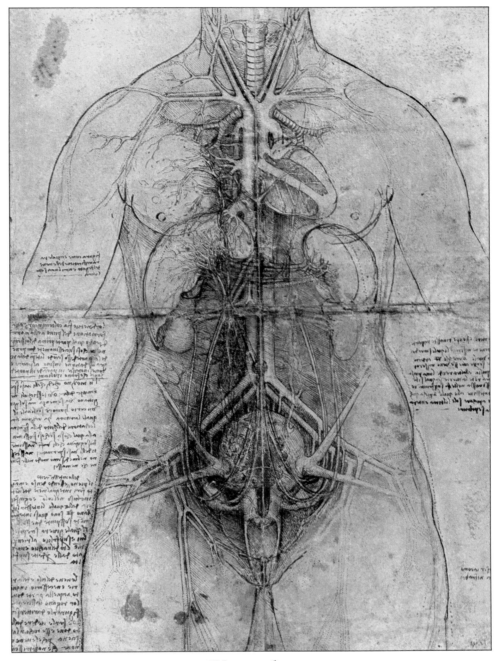

Written by
EMILY BLAIR CHEWNING
Designed by
DANA LEVY

A Fireside Book
Published by
Simon and Schuster New York

For Richard
and for Dr. E. P. Parker III

A special thanks to Linda Sunshine for
her suggestions and encouragement
throughout.

Published by Simon and Schuster
A division of Gulf & Western Corporation
Simon & Schuster Building
Rockefeller Center
1230 Avenue of the Americas
New York, New York 10020

Designed by Dana Levy

Manufactured in the United States of America

1   2   3   4   5   6   7   8   9   10

Library of Congress Cataloging in Publication Data

Chewning, Emily Blair.
  Anatomy illustrated.

  (A Fireside book)
  1. Anatomy, Human — History — Pictorial works.
2. Anatomy, Artistic — Pictorial works.   I. Levy,
Dana.   II. Title.
QM25.C46   611'.022'2   78-24548

ISBN 0671-24597-X

Cover illustrations: A university dissection scene: 1491
Fasciculus Medicinae Johannes De Ketham
Yale Medical Library
A color-coded x-ray of a healthy woman wearing a
necklace: 1970s Agfa-Gevaert, West Germany

Title page illustration: Anatomical chart of the principal
organs and arterial system of the female body: late
15th century Leonardo da Vinci
Windsor, Royal Library (12281 recto)

Contents page illustration: The proportions
of man: 1st century B.C. Lucio di Vitruvio
Yale Medical Library

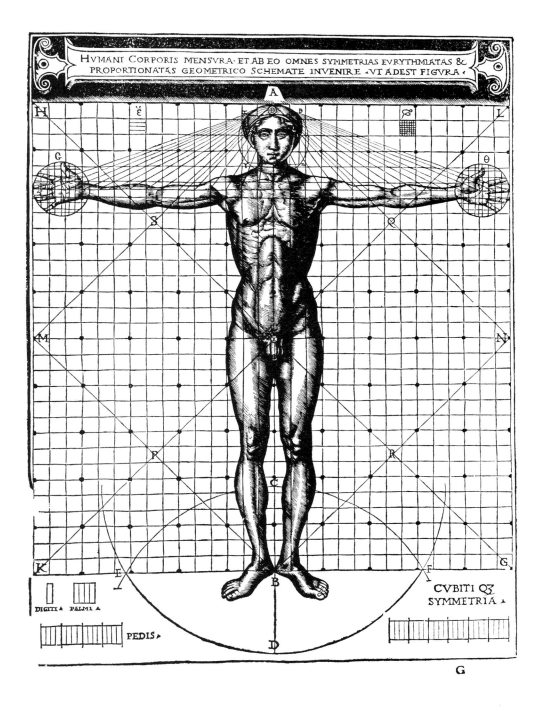

HVMANI CORPORIS MENSVRA·ET AB EO OMNES SYMMETRIAS EVRYTHMIATAS &
PROPORTIONATAS GEOMETRICO SCHEMATE INVENIRE ·VT ADEST FIGVRA·

DIGITI   PALMI

PEDIS

CVBITI Q3
SYMMETRIA

G

Contents

Introduction

It is wrong to think of science as a mechanical record of facts, and it is wrong to think of the arts as remote and private fancies. What makes each human, what makes them universal is the stamp of the creative mind.
Jacob Bronowski, *Science and Human Values*

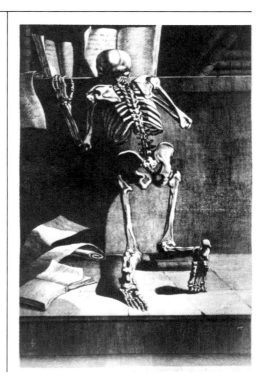

From primitive man's first crude sketches to the sophisticated images of today's science, the complex mystery of the human body has fired man's imagination. Science has explored the human body, searching for clues to the origin of life itself. Art has probed the human figure, investigating the intricate structure of its form. Whether through the eyes of science or the mind of art, the same essentially creative force has inspired a wealth of anatomical observations.

Although the ancient cultures of the world produced exquisite art and sculpture of the human form, the first scientific studies of the human body appeared during the Middle Ages. The study of gross anatomy, that is observation based upon dissection of the body, inspired intricate woodcuts and engravings that show the body in great detail. The growing emphasis on accuracy and detail culminated in tremendous progress for the study of the human form during the fifteenth century.

The Renaissance produced the genius of the artist in Leonardo da Vinci and the genius of the scientist in Andreas Vesalius. Each of them independently produced monumental studies of the human form. Although only a few years separated them between the fifteenth and sixteenth centuries, they never knew each other's work.

ABOVE: Engraving of a skeleton, eighteenth century. Jacques Gamelin. Library of the University of Toulouse.
OPPOSITE: A skeletal system from a Persian manuscript, c. 1400. Reproduced by permission, Director of the India Office, Library and Records.

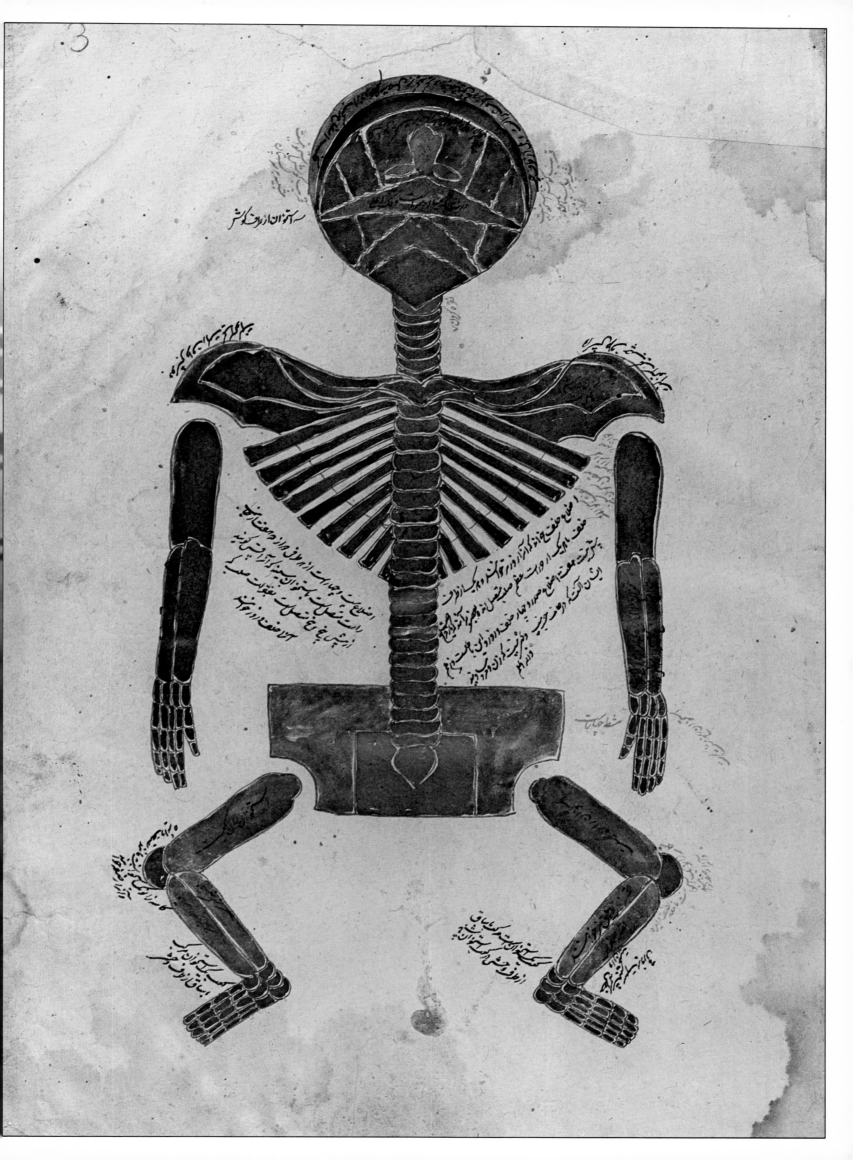

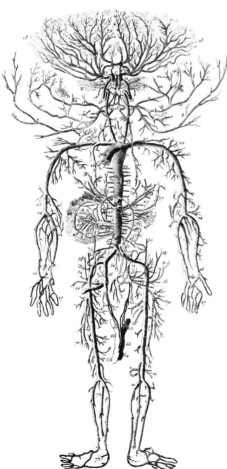

OPPOSITE: *The Anatomy Lesson*, mid-seventeenth
century, by Thomas de Keyser. Amsterdam
Historical Museum. LEFT: An engraving of the
arterial system from the Encyclopedia of Diderot
and D'Alembert, 1745. Bibliothèque Nationale,
Paris. ABOVE: *An Anatomy Lesson (The Reward
of Cruelty)*, 1759, by William Hogarth. Bibliothèque
Nationale, Paris.

Leonardo, whose work was completed first, spent his lifetime studying the human body. His observations, which were collected into notebooks and unfortunately lost shortly after he died, not to be discovered again until the twentieth century, were the first real scientific studies of the human body in history. Leonardo's method of systematic observation set the standard upon which all modern science is based.

Andreas Vesalius' masterwork of human anatomy was not done, as was Leonardo's, over a lifetime. He completed his work, including his masterpiece, *The Seven Books on the Structure of the Human Body*, by the age of twenty-eight, compiling, by that time, the foundations of the modern science of human anatomy. His beautiful illustrations of the human body represented the most thorough and accurate studies ever organized into a single work. Vesalius was solely responsible for what has been dubbed "the bible" of human anatomy.

Leonardo and Vesalius were explorers in the unknown territory of the human form. They were driven by a passion for knowledge. They pursued the unknown relentlessly and made it the known. They were also both deeply concerned about the precarious nature of human existence. The more they could observe man, the more they could hope to understand him. It seems irrelevant to label one artist and the other scientist. Clearly, the inventions of science and the inspirations of art flow out of the same creative pro-

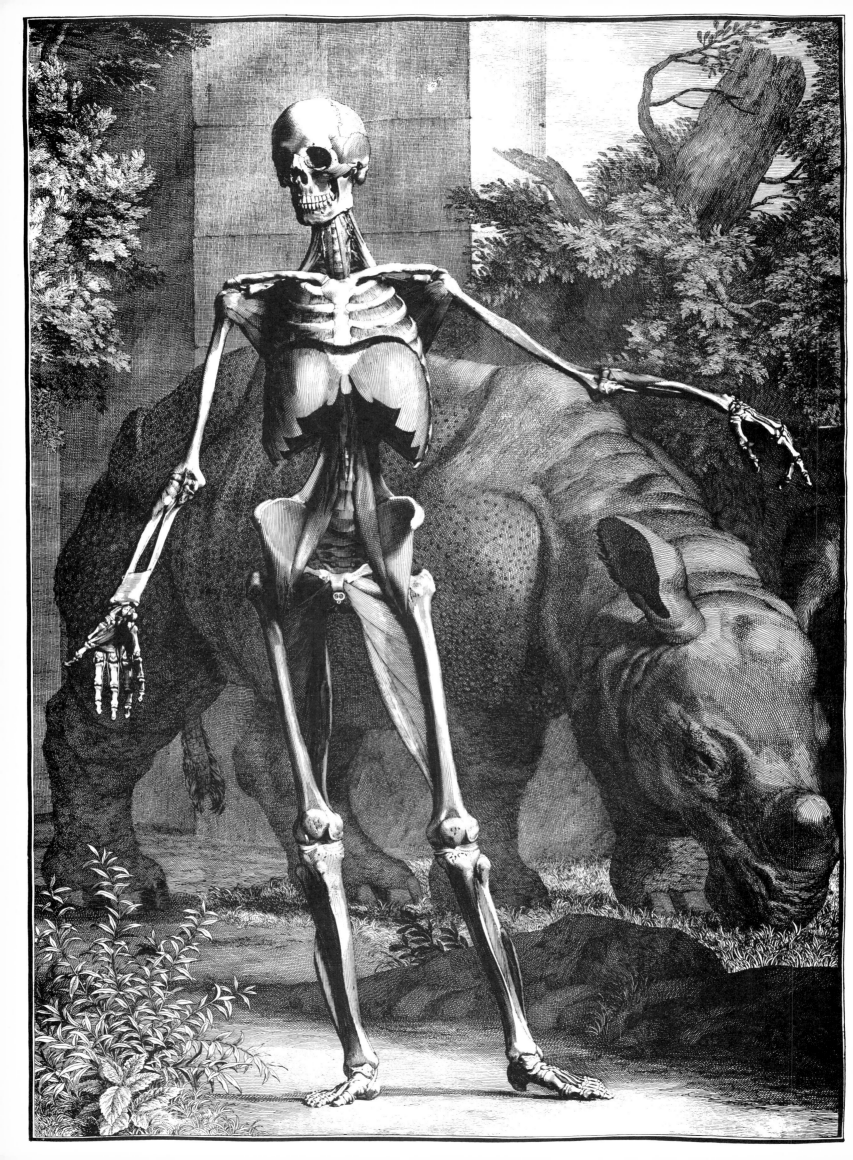

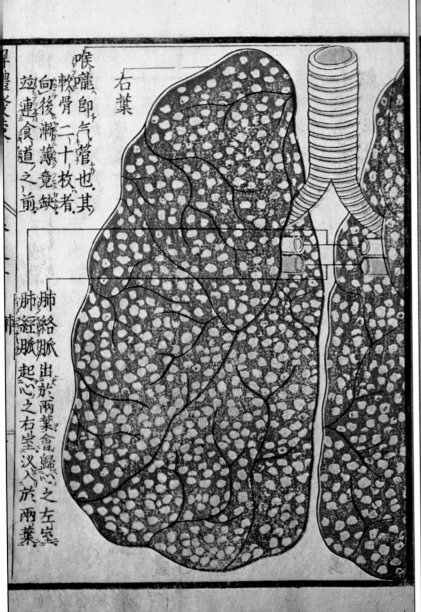

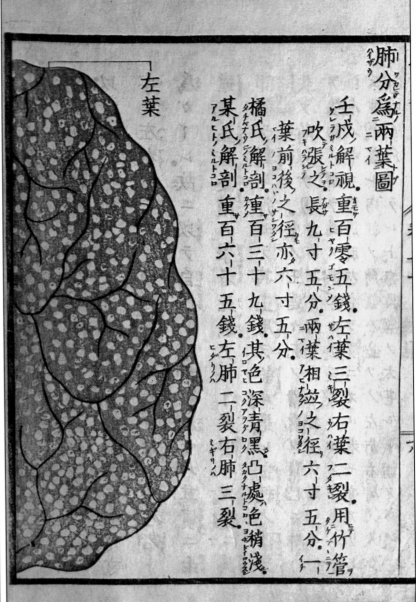

cess. These men were universal spirits whose contributions to the field of human anatomy will endure as milestones in the history of the world.

The dazzling anatomical illustrations that followed in the wake of Leonardo and Vesalius grew out of creative efforts in the fields of both science and art. The sixteenth and early seventeenth centuries saw artistic anatomies produced that were as faithful to scientific detail as scientific anatomies were sensitive to an artistic ideal. Beautiful atlases of the human form were published during these years, and scientific illustrators began to experiment with color.

From the end of the seventeenth century to the present day, modern science's extraordinary new imaging techniques have awed the world. The discovery of the light microscope in the seventeenth century brought the

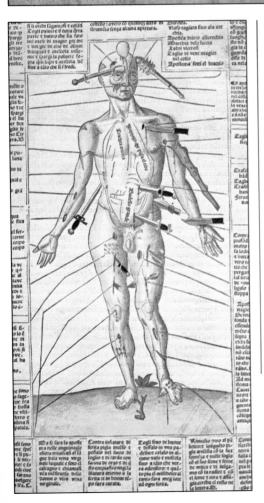

OPPOSITE: The proportions of the human skeleton compared to the figure of a rhinoceros, 1747, by Bernhard Siegfried Albinus. Yale Medical Library. ABOVE: A Japanese woodcut print of the lungs, late eighteenth century, artist unknown. Courtesy of Gordon Mestler. LEFT: *The Wound Man* from Johannes De Ketham's *Fasciculus Medicinae*, 1491. Yale Medical Library. *The Wound Man* was used as a model to teach the treatment of wounds by physicians in the Middle Ages.

invisible world of the human cell into
focus and gave rise to the study of
cellular anatomy. The subsequent
invention of the electron microscope
further exposed the tiny world of the
molecule, and the study of molecular
anatomy was born.

Modern science in the twentieth
century has produced the ultimate
technological portrait of the human
form. The latest imaging techniques
use x-rays and sonar to scan the
interior of the human body, revealing
strange and beautiful sights that
stagger the imagination. Recent
advances in optical systems have made
it possible for man to photograph
areas of the body that were inacces-
sible to the naked eye until just a few
years ago.

Man's exploration of man has always
touched a most human nerve at the
core of our being. Whether the

initiative belongs to the artist or to
the scientist, the inspiration is human-
ism. Man is endlessly complex, endlessly interesting. To delve into the
study of the human form is to delve
into the study of man. History's record
of man's observations on the human
form attest to the richness and remarkable variety of the experience.

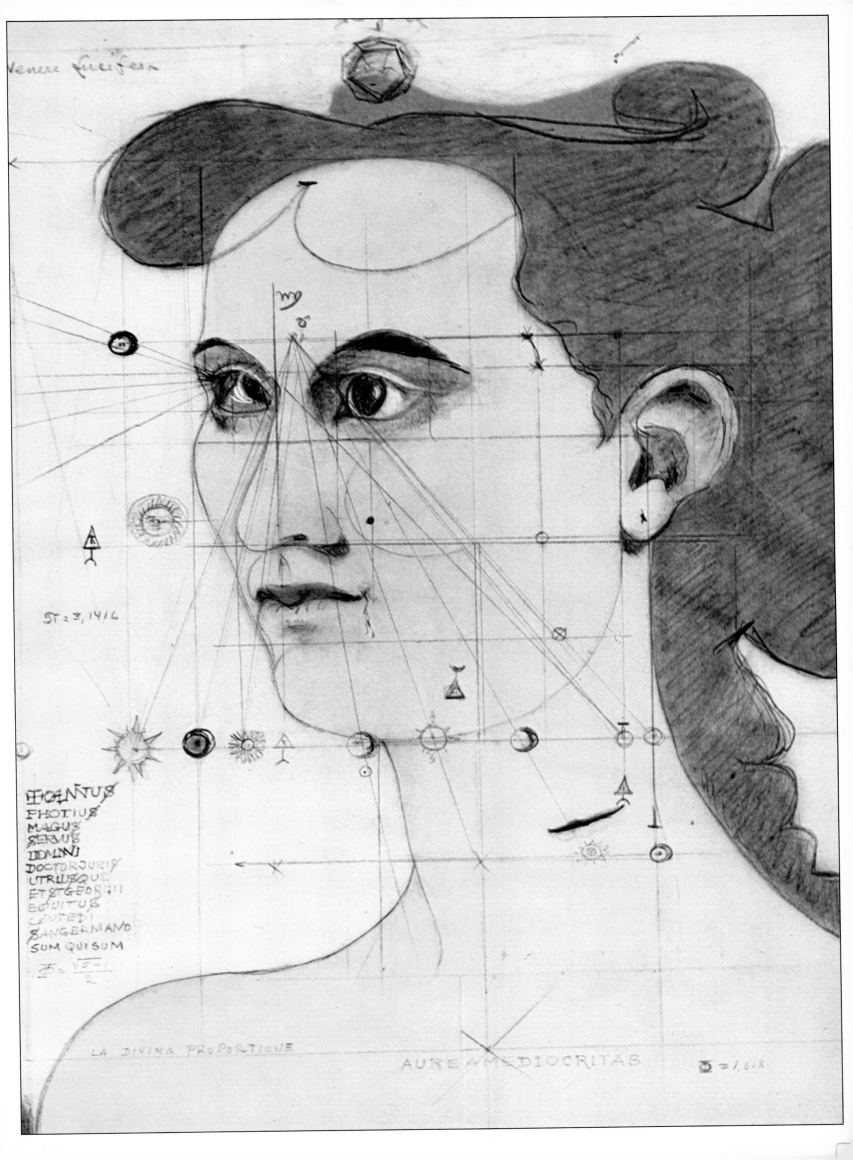

OPPOSITE: Perspectives of the human figure, mid-seventeenth century, by Chrisóstomo Martínez. The New York Academy of Medicine.

ANATOMY
ILLUSTRATED

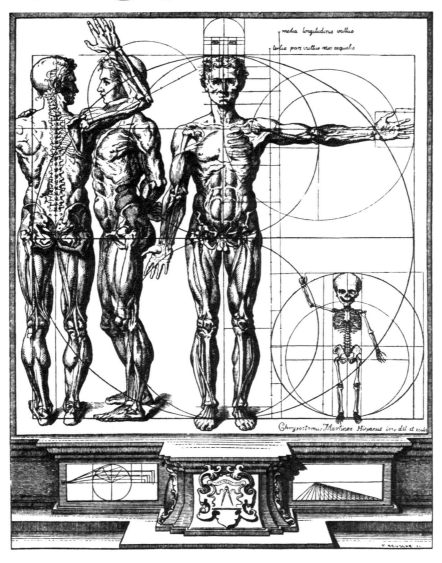

The Zodiacal Man

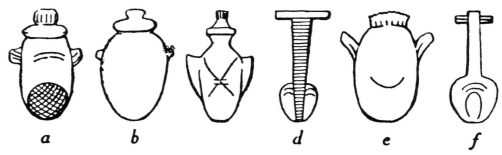

a   b   d   e   f

Although the study of human anatomy didn't really begin until the time of the Renaissance, the human figure has always fascinated man. Crude attempts at rendering the human form were discovered in the caves of primitive man. In their hieroglyphics and in their surgical papyri, the ancient Egyptians made many references to the parts of the body. The Greeks and the Romans produced exquisite sculpture which reflected their genuine obsession with the physical beauty of the human body.

Primitive man's rendering of the human figure indicates that he understood the basics of anatomy. For example, he recognized the location of the heart, and he represented death by drawing an arrow puncturing the heart. In a few rare instances, there are also some sketches in which the arms and legs are bent at the joints.

The ancient Egyptians demonstrated, for their period in history, a sophisticated knowledge of the human body. While graphic images are somewhat scarce, there are numerous descriptions of different parts of the body in their writing. In fact, there are descriptions of a manual on anatomy that is supposed to have been written by a king-physician as early as the first dynasty.

Because of their practice of embalming the body at death, the Egyptians had frequent opportunities to observe the structure of the body. They certainly knew something of the heart and the circulation of the blood, the location of the liver and the intestines, and they knew something about the brain. By the year 350 B.C., the Alexandrian scholar Herophilus had finished his work on human anatomy, which for the first time recognized the sections of the brain.

The anatomical observations of the Greeks and the Romans found expression in their sculpture. They rendered the surface contours of the body in intricate detail. Their figures display surprisingly accurate bone structure and muscle

systems. The Greeks and the Romans did a great deal to advance anatomical theory in general. It was certainly their efforts which established the basis for the progress of anatomy during the Renaissance. Yet during these early stages, it was their translation of anatomical observation into art which inspired interest in the study of anatomy.

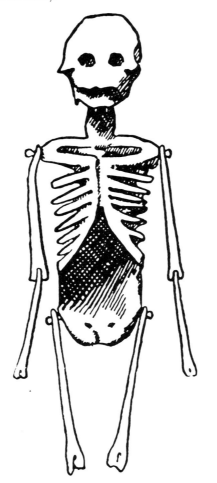

Only when the Dark Ages settled over Europe in the first century after Christ did man's interest in the human figure wane. The growing power of the Christian Church discouraged the pursuit of all worldly knowledge. The Church was especially critical of any attempt to study the body. The soul, it maintained, should be the main focus of man's attention; it would have been heresy to raise the body to the level of the soul by placing any emphasis on its importance.

Because of Christianity's power over Europe, medieval man fell back on the mystique of astrology to represent the human form. Probably the most common diagram drawn during the tenth and eleventh centuries is that of the Zodiacal Man. In this diagram, the different parts of the body are identified with certain stars which were said to influence the functioning and well-being of the individual. The Church was able to accept this explanation of the body: after all, the existence of the heavens and a system of stars only served to confirm the existence of the even higher authority of God.

OPPOSITE: "The Zodiacal Man" from Johannes De Ketham's *Fasciculus Medicinae,* 1491. Yale Medical Library. ABOVE: Egyptian hieroglyphic representations of the heart (a,b,c,e), the trachea and the lungs (d,f), c. 2900 B.C. LEFT: Ancient Roman representation of the skelton, c. 50 B.C.

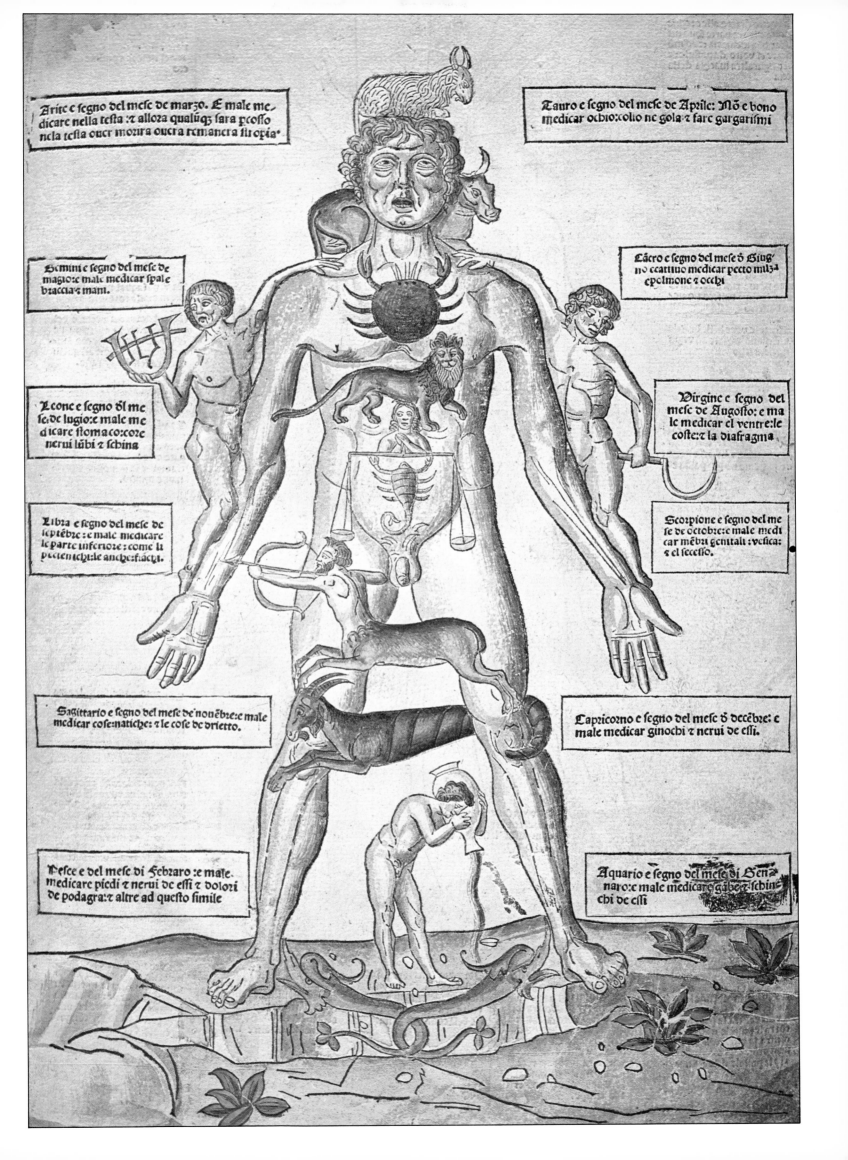

Arite e segno del mese de marzo. E male me-dicare nella testa: e alloza qualunqz sara pcosso ncla testa ouer mozira ouera remancra stiropia.

Tauro e segno del mese de Aprile: Mo e bono medicar ochio: collo ne gola e fare gargarismi

Hemini e segno del mese de magio: e male medicar spale braccia e mani.

Cancro e segno del mese d Giugno e cattiuo medicar pecto milza epolmone e occhi

Leone e segno dl mese de lugio: e male me dicare stomaco: core nerui lūbi e schina

Virgine e segno del mese de Augosto: e male medicar el ventre: le coste: e la diafragma

Libra e segno del mese de septēbre: e male medicare le parte inferiore: come li pectenichi: le anche: fiachi.

Scorpione e segno del mese de octobre: e male medicar mēbri genitali: vesica: e el secesso.

Sagittario e segno del mese de nonēbre: e male medicar cose: natiche: e le cose de drietto.

Capricorno e segno del mese d decēbre: e male medicar ginochi e nerui de essi.

Pesce e del mese di Febraro: e male medicare piedi e nerui de essi e dolori de podagra: e altre ad questo simile

Aquario e segno del mese di Bennaro: e male medicare gābe e schinchi de essi

Medieval Images

Anatomical images began appearing in Europe around the thirteenth century in the medical manuscripts of physicians. Until this time, there had been few formal attempts to represent the structure of the human body. The artwork of ancient Egypt, Greece and Rome contains beautiful drawings and sculptures of the human figure, but the images only loosely refer to the actual structure of the body. The early years of the Middle Ages produced only the Christian Church's wrath against the idea that there was anything to learn about human anatomy. Only when these Dark Ages drew to a close and Europe awoke with a passion for knowledge was the time right for the widespread circulation of anatomical figures.

As stunning as these medieval medical manuscripts were, the actual drawings of the human form were still at a very

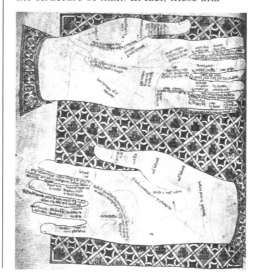

primitive stage. Because the Church forbade dissection, there were few opportunities to observe the details of the structure of man. In fact, these anatomical drawings were merely schematic representations of the body and a few of its main parts.

Medieval illustrators worked from a stylized format using the same model that had been handed down from the ancient world. They generally did a series of five or six drawings, each representing a different system of the body; there were the osseous, the nervous, the muscular, the venous and the arterial systems. On occasion, they would include a sixth drawing of the reproductive system as seen in the pregnant woman. All of these figures are drawn in squatting positions with the arms above the head.

These schematic diagrams were not done, at this point in time, for the specific purpose of studying the human structure. Rather, physicians used these diagrams as models to demonstrate medical techniques for their students.

OPPOSITE: An early medieval schematic diagram of the muscle system, early fourteenth century. Ashmolean Codex 399. Ashmolean Museum, Oxford. ABOVE: A chiromantic drawing of the hands. Chiromancy involves the art of divination through analyzing the structure and appearance of the hands.

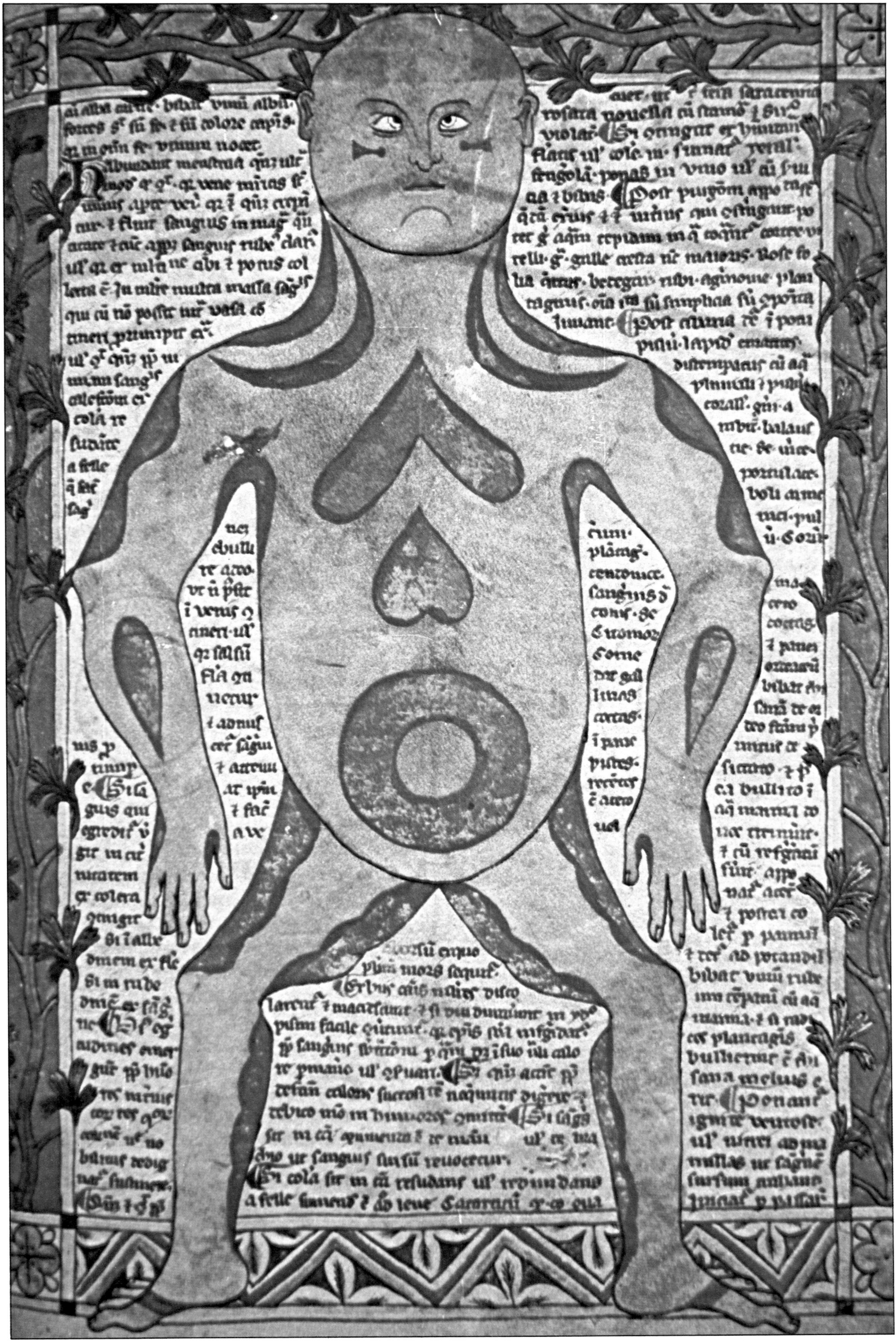

The First Human Dissections

The internal structure of man remained largely unknown until the practice of human dissection began at the end of the thirteenth century. It wasn't until physicians were allowed to do dissections on human cadavers, that the study of anatomy could progress beyond the surface observations of the past. With the exception of the Egyptians, who practiced embalming, there simply had been no access to the human body. Early anatomical images became stylized into antiquated, schematic diagrams. In fact, it was only the artists during this period who had any interest in rendering even the surface of the human body in detail.

The first dissections of human cadavers took place in the universities of Europe. These dissections were done not for the

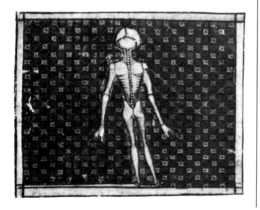

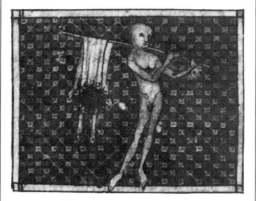

purpose of observation, but because the law schools of the day needed autopsies performed to determine criminal intent. The physicians recorded these events, but the emphasis lay more on the methods of dissection than on the structure of the human body.

It was actually the work of a surgeon, William of Saliceto (1215-1280), from the University of Bologna, which contained the first official instances of human dissections' being performed and recorded. While there had been scattered references to a random dissection, William of Saliceto's

treatise (which he finished in 1266) provided the first solid evidence of the practice.

Several diagrams done in the early fourteenth century by another University of Bologna surgeon, Henri de Mondeville (1270-1320), showed some of the techniques that were used in the dissection procedure at that time. He was a Norman student who studied and taught at the University of Bologna. He later returned in 1304 to teach at his own University of Montpellier. He probably had the diagrams done while he was at Bologna and carried them back with him when he went to teach in his native country.

De Mondeville, like the rest of his contemporaries, still focused on methods and techniques of dissection rather than on the study of the structure of the body. However, the practice of dissection opened the way for the great work in anatomy that was to come.

OPPOSITE: An anatomical miniature from the French manuscript of Henri de Mondeville, 1314. ABOVE: Three other miniatures from the same French manuscript. All are from the Bibliothèque Nationale, Paris.

ek et de la face et comment il sent
presentent a ceus qui tel regar
dent de coste. ~ ~ ~ ~

At trues ci noter que que dit le
commun de la diuersite des os
et des commissures du chief de saiute
tdoine et iasoit ce que le philoso

Medieval Woman and Anatomy

Before the Renaissance, the traditional approach to the anatomy of woman was to depict her during pregnancy. The age-old emphasis on propriety was too powerful to permit the representation of her unclothed body in the interest of science alone. The image of the pregnant woman, however, was the exception. It reflected the importance to society of her responsibility as childbearer, the carrier of the next generation. The study of the reproductive system seemed to justify the study of the female form.

The schematic model used for the pregnant female figure was passed from the ancient world right up through the Middle Ages. The woman was always depicted squatting with her arms raised above her head. Rarely were any of the external female organs detailed. Occasionally figures were drawn with breasts, but this was not common. In addition, the uterus tended to be shaped like a lightbulb and was contained inside another structure of a similar shape,

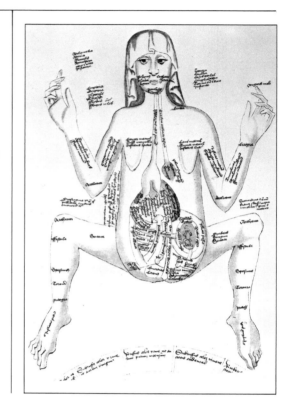

probably the stomach. The fetus was usually sketched with its hands over its eyes and at full-term development. In those days, it was generally believed that the fetus was conceived fully formed. The liver had been located, but it was without lobes, and the heart was encased inside the lungs.

There were few female cadavers for dissection, and thus few chances to observe the female form in detail. The supply of cadavers was limited to those of executed criminals, and it appeared that there were more men than women sentenced to die for criminal acts. Consequently, it would have been even more difficult to come upon a pregnant female cadaver.

There were scattered references in surgical manuscripts to caesarean operations to remove the fetus from the womb in cases of problem deliveries. Since few women ever survived this procedure, there was some chance to observe their internal structure at this time.

The study of the anatomy of woman fell behind that of man until the Renaissance, when Leonardo da Vinci did his beautifully detailed studies of the female form.

OPPOSITE: *Gravida* from a miniature painted c. 1400. Leipzig Manuscript Codex 1122. Yale Medical Library. ABOVE: A schematic diagram of a pregnant woman from Johannes De Ketham's *Fasciculus Medicinae*, 1491. Yale Medical Library.

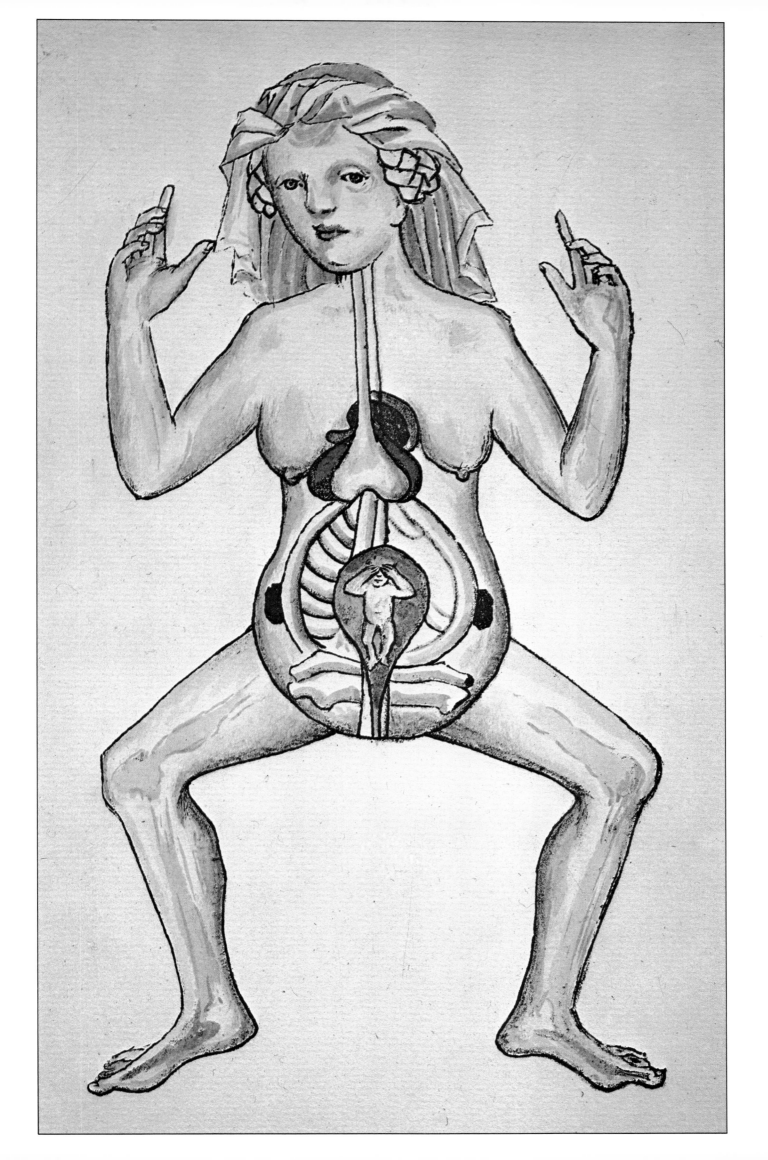

Anatomy in the Medieval University

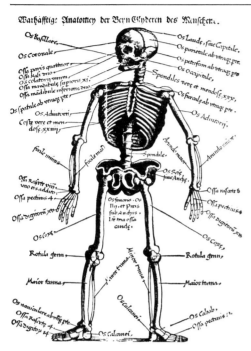

By the end of the 1400s, the study of human anatomy was officially recognized as a part of the curriculum in universities all over Europe. Dissections of human cadavers, which had been performed solely for the purpose of autopsy, suddenly were routinely permitted for anatomical observation. Professors and students gathered around the cadavers, which were stretched out on the cold marble slabs, to observe the totality of the human form. Curiously, it was a period in which the study of anatomy made very few advances. It was as if the weight of scholarly recognition had sapped a certain initiative from the energy of pure observation.

What happened to the practice of dissection when the university finally acknowledged human anatomy in its curriculum was symptomatic of the character of the academic world at that time. The moment anatomy won the approval of the university, the professors' stature outgrew the subject matter. Instead of handling the dissections themselves, these professors ritualized the procedure into a performance. They removed themselves from the scene of the cadaver to a thronelike chair positioned high up on a podium from which they could lecture to their students. They hired two assistants to do the actual work. One assistant made the incisions and handled the body. His was considered a menial task; very often, this individual was a butcher in everyday life. The other assistant was usually a younger colleague. He was supposed to point with his special wand to the incisions as the professor lectured. His title was "ostensor." With its formalized proceedings, the atmosphere of the university was certainly not one to inspire new research or discovery.

The study of anatomy was to continue its plodding course in the university for many years. The real source of inspiration for the reform of human anatomy would come from the imagination of an artist, not from the pragmatic rhetoric of a scholar.

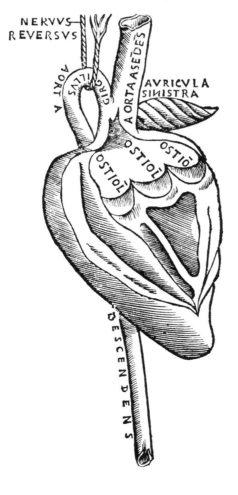

OPPOSITE: A university dissection scene, 1491, *Fasciculus Medicinae*, by Johannes De Ketham. Yale Medical Library. ABOVE, LEFT: A German woodcut of a skeleton and its various parts, sixteenth century, artist unknown. Yale Medical Library. LEFT: Berengar of Carpi's drawing of the heart, done while at the University of Bologna, 1532. Yale Medical Library.

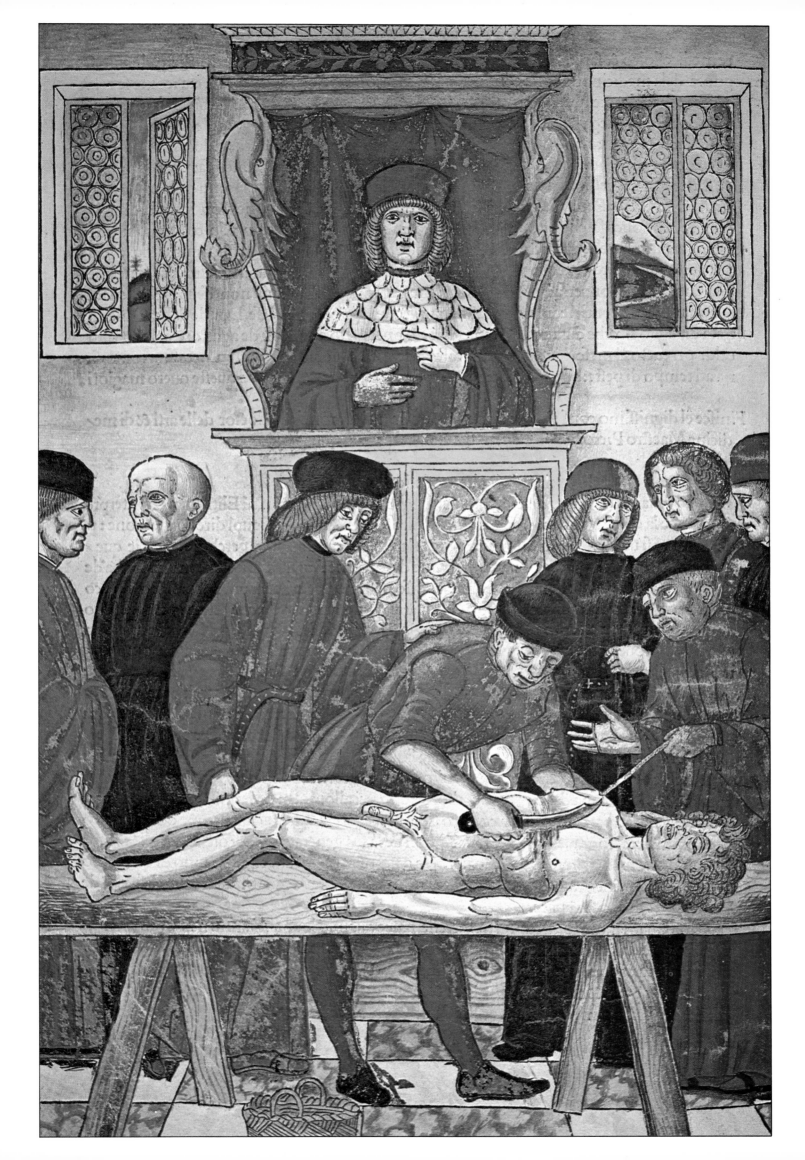

Leonardo da Vinci

"The nerves with their muscles serve the tendons even as soldiers serve their leaders and the tendons serve the common sense as the leaders their captain and the common sense serves the soul as the captain serves his Lord." Leonardo da Vinci (Anatomy Ms. B, Windsor 19019)

Leonardo da Vinci's (1452-1519) contribution to the study of human anatomy was awesome. With his artist's eye for detail, he explored the unknown territory of the human body as it had never been done before. He probed the body with rare foresight, making anatomical observations that were far in advance of their time.

Leonardo was obsessed with detail. He believed it was the detail in nature that gave meaning to the Grand Design and, thus, the detail of the human structure that gave meaning to the whole of man's nature. Although it was as an artist that he struggled to understand the nature of man, it was as a scientist that he undertook the investigation of man.

Leonardo kept anatomical notebooks packed with his observations and experiments from around 1489 right up until the time of his death. As a young man, he managed to convince the town surgeons of his interest in the structure of the

human form, and thereafter he regularly attended their dissections. Leonardo had planned to publish a textbook on anatomy with the Pavian professor Marcantonio della Torre, but the professor's untimely death at the age of thirty-one halted their plans. When Leonardo died, his notebooks were lost, and they were not found again until the twentieth century. Had those notebooks been available to the world at the time of Leonardo's death, the evolution of anatomy would have been speeded up by many years. These notebooks contained material on the human body that would have proved

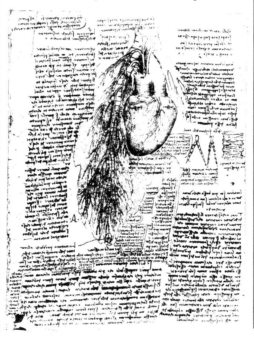

invaluable to the anatomists working at the time.

In one of his manuscripts, Leonardo set down the outline for his anatomical textbook. He was to have produced nine diagrams for bones and muscles, three diagrams of the nervous system and three of blood circulation, as well as six cross-sections of man and woman. Leonardo was the first to draw the parts of the body from different angles—the front, back and side views. He was also the first to show cross-sections of the structure of the body. Both these techniques are used by anatomists today.

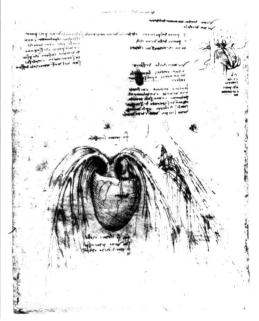

OPPOSITE: A study of skulls, late fifteenth or early sixteenth century, by Leonardo da Vinci. Windsor, Royal Library (19057, Recto). ABOVE: Leonardo's studies of the heart, late fifteenth or early sixteenth century. Yale Medical Library.

io vogho levare quella parte d'losso della armatura della ghuaera
che sta nova infralle 4 line a b c d e pla scopri aprira dimo
strare la largeza eprofondita d' 2 baghi e d' drieto aquello lasse godano
nel vacquo d'sopa lassedori teguostrupere d'la vsste indsis llo disoto ista eme
re notritore delle redie d't denti

jl vacquo d'losso d'ghuaera assimilitudine pprofondita eplargeza
del vacquo d'ricicui d'mo losgra asf losgio e pgapacira e molto
semple aosso ericicui d'mo asf nini phlusi m iquah d'sse aruno
fabe aruello passanto plo gosasorio d'ssesania la supfrusta delli omori
d'la testa nel naso altri dusi custerno similinda alui d'l vacquo d'sopa
esorrazusa losgio ilbuso b edobsiamicu di una passa allso ilbuso
m egosfeberi simso talegori acosgio passanto plo gamal d'l naso

Leonardo's diagrams of the parts of the body are accurate down to the finest points. His studies of muscles are remarkable. He had been sketching them on models since he was a young boy. Once he was able to observe them beneath the surface of the body, he was able to render their every detail. His studies of the heart show quite a knowledge of its structure. He experimentally determined the movements of the heart, and he built models of the valves to show their action. Leonardo performed the first anatomical injections to solidify the cavities of the brain, taking casts of them from which he was able to study its structure. He made the first complex models of the skeleton and its various parts. Finally, Leonardo's studies of the womb and the fetus provided the first dramatic look at the reproductive system of woman.

Although Leonardo's notebooks were lost, he left the world his other work, especially his paintings, which were invested with the same quality he had put into his anatomical studies. He was a man whose vision transcended the confines of both art and science. He was an inventor and a creator of universal images. Leonardo's observations on the structure of the human form were one part investigation and one part inspiration. His vision was timeless, and its impact is as fresh and powerful today as it was during his lifetime centuries ago.

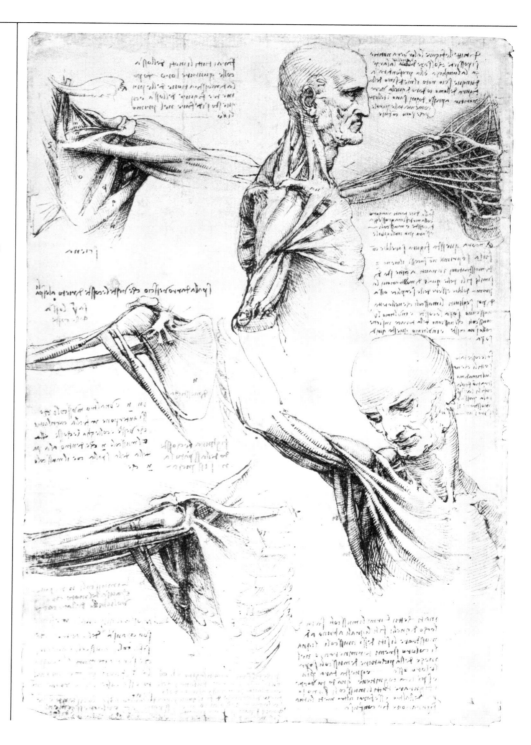

OPPOSITE: Studies of the fetus, late fifteenth or early sixteenth century, by Leonardo da Vinci. Windsor, Royal Library (19102, Recto). ABOVE: Anatomical muscle studies, late fifteenth or early sixteenth century, by Leonardo da Vinci. Windsor, Royal Library (19003, Recto).

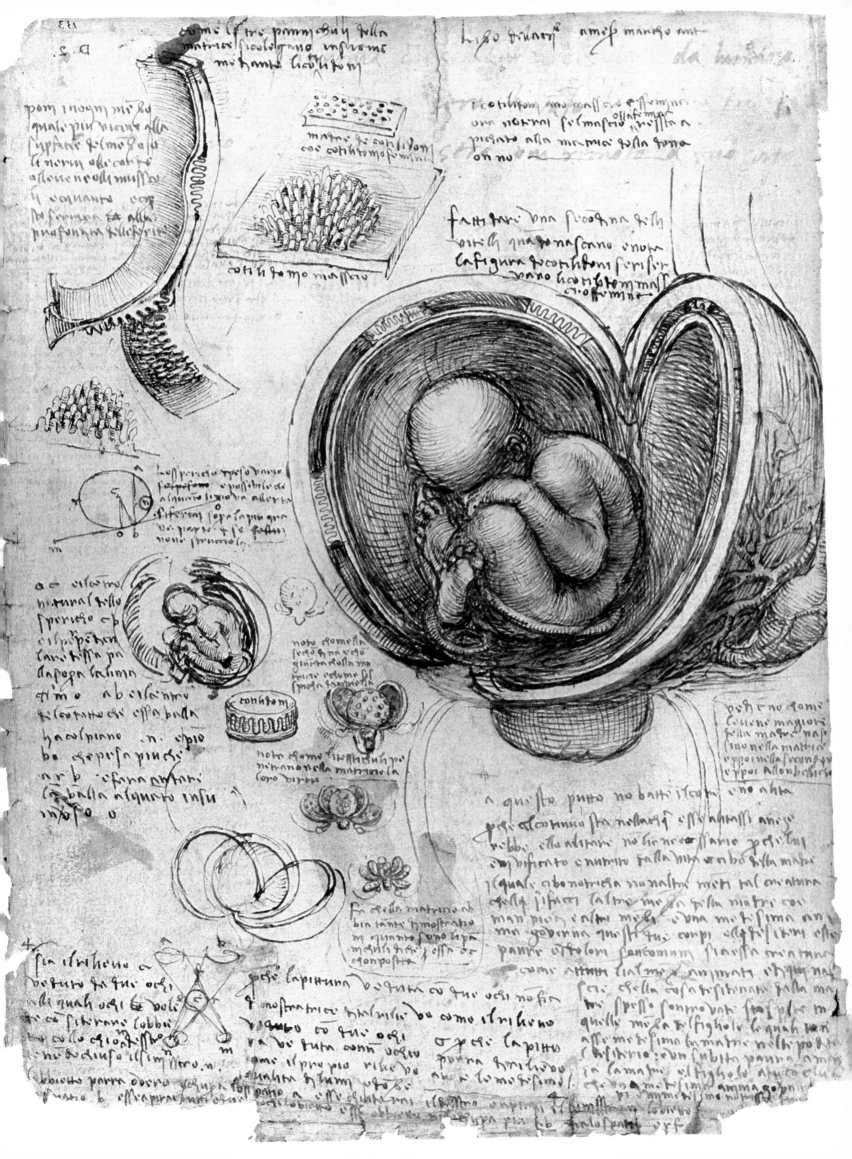

Andreas Vesalius

The considerable reforms of Andreas Vesalius (1514-1564) in the field of human anatomy not only laid the foundations of modern medicine; they also marked the first significant achievement in modern science itself. He undertook the most thorough and accurate study of the human form ever attempted. Because he was also a product of the Renaissance and, as such, an artist in his own right, his observations were collected into volumes which exquisitely illustrated the dimensions of man. His most famous work, published in 1543, was titled *De Humani Corporis Fabrica Libri Septem* (Seven Books on the Structure of the Human Body).

Vesalius was born into a family of physicians in 1514 in Brussels. He is known to have been intrigued with the structure of animals and was already dissecting them for study at a very young age. He was sent to Paris as a young man to train with some of the best men in the medical field. By the age of twenty-four, he had already produced a work of his own observations called *Tabulae sex*. At this point in his life, although his observations were clearly presented, Vesalius was still using the diagrams of old anatomies.

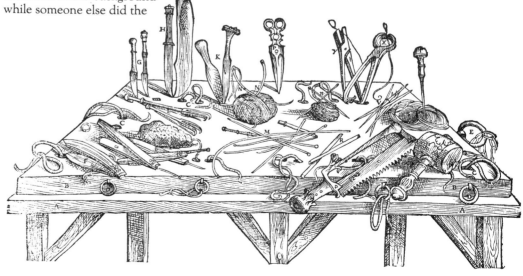

It wasn't until he had traveled to Italy and been appointed professor at the University of Padua that Vesalius' real work in anatomy would begin. His first act at Padua was to do away with the "ostensors." Those were the individuals who pointed with their wands to the incisions made during the dissection procedures while the professors sat back in their high chairs and lectured. Vesalius did not want to remain in the background while someone else did the active work of the dissection. He got down off his "throne," did away with the hired help and, with his students, went to work. He knew he would never make any observations from the professor's podium. Vesalius also initiated the use of live models with the outlines of different parts of the body drawn right on their skin.

During this time, Vesalius is said to have had enormous energy and a steely determination to achieve his purpose. It must have been true, because he completed the major work of his life, *De Humani Corporis Fabrica*, by the time he was twenty-eight years old.

All the anatomical observations he made while professor at Padua were recorded and arranged in his *Fabrica*. (He also published a similar volume, *Epitome*, around the same time which was intended for the lay public.) *Fabrica* is divided into seven sections, describing the various systems of the body. Vesalius' observations on muscles, bones, joints and the nervous system are superior to his descriptions of internal organs. Nevertheless, the entire collection of his work was far in advance of anything that had ever been done before. *Fabrica* also represented the first serious attempt in any discipline to gather so much observation into a single work.

OPPOSITE: The muscle system, 1543, by Andreas Vesalius, *De Humani Corporis Fabrica Libri Septem*. Yale Medical Library. ABOVE: Portrait of Andreas Vesalius from *Fabrica*. RIGHT: Vesalius' dissection tools, also from *Fabrica*. Both are from the Yale Medical Library.

The accuracy that Vesalius achieved in his illustrations was astounding. Like the other anatomists of the day, Vesalius had little time to observe the details of the structure of the body during dissection. Because there were no preservatives, human cadavers began to decompose more quickly than the anatomists could work on them. Yet despite these difficulties, Vesalius skillfully re-created in *Fabrica* the structure of man.

Vesalius' illustrations of the human body are beautiful. But then, he was an artist, and he considered the human body God's most beautiful creation. Despite the seven-section scheme of *Fabrica* he did not see the parts of the body as being separate and as having independent functions the way anatomists of today do. He was interested in the whole "fabric" of the body. His drawings are composed of figures in lifelike positions, set against the background of everyday life. Anatomy of that period meant living anatomy, in the belief that the only way to observe the human body was in its natural, alive state.

Finally, Andreas Vesalius was different from the other great anatomists in history who chose to attach their names to a special part of the body. Anatomy has the tube of Fallopius, the canal of Eustachius, the veins of Galen and many more parts named for those who described them. Vesalius became known instead for his work on the whole body. Vesalius was not preoccupied with the ideal of the human form, he was intrigued by the whole miraculous design of man.

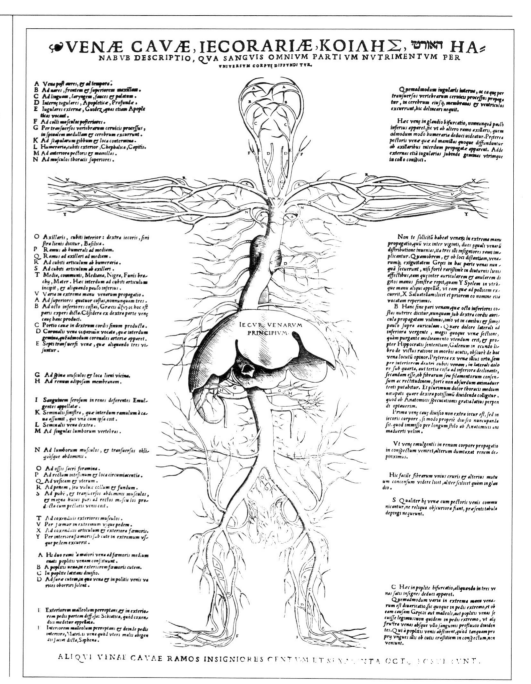

OPPOSITE: A detail from a Diego Rivera mural of Andreas Vesalius, 1943. Courtesy of the Instituto Nacional de Cardiologia, University of Mexico. ABOVE: The venal system, 1543, by Andreas Vesalius, *De Humani Corporis Fabrica Libri Septem.* The New York Academy of Medicine.

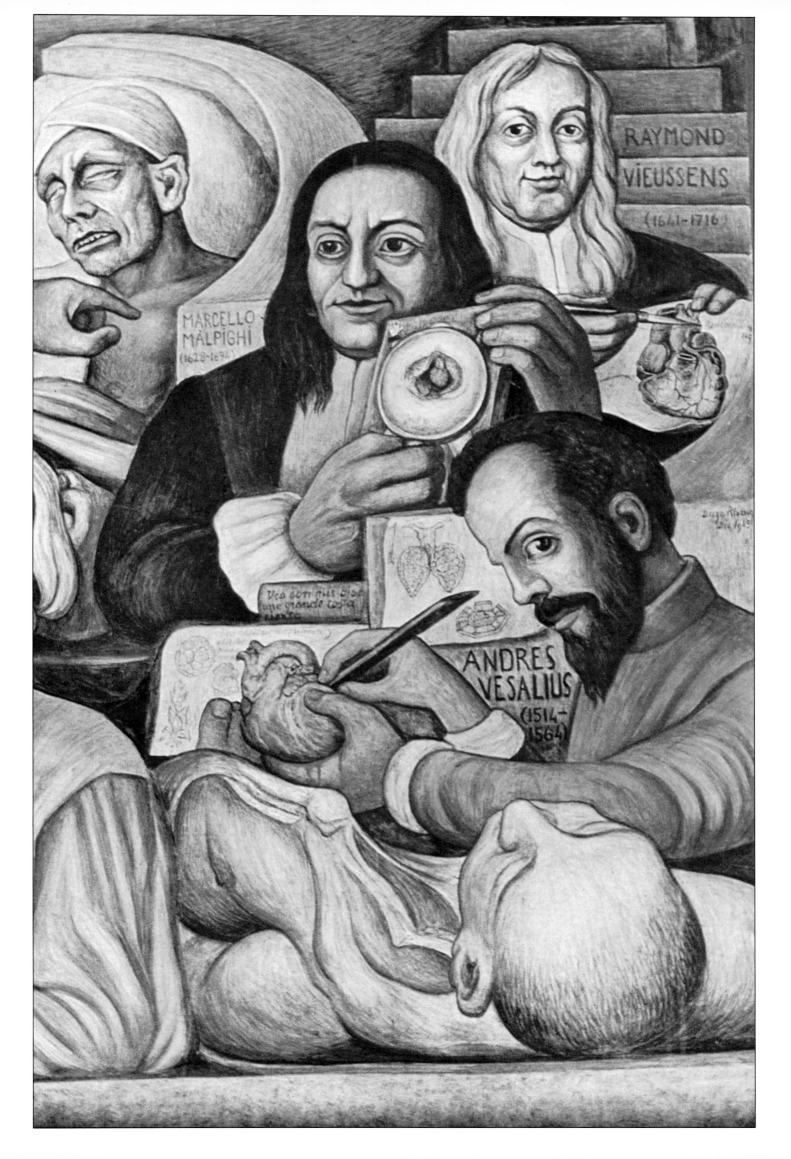

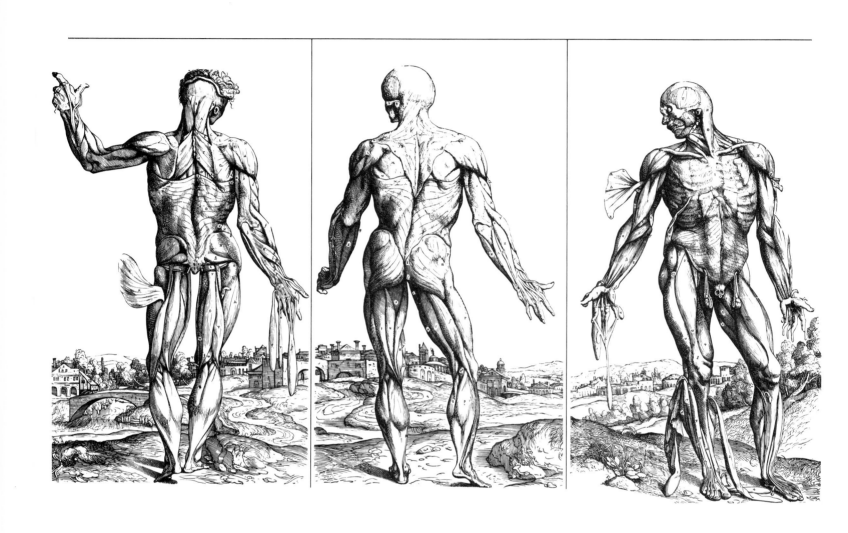

OPPOSITE: The skeletal system. ABOVE: Three
views of the muscles and tendons. All four plates
are from *De Humani Corporis Fabrica Libri Septem*,
1543, by Andreas Vesalius. Yale Medical Library.

CORPORIS HVMANI OSSA
POSTERIORI *FACIE PROPOSITA.*

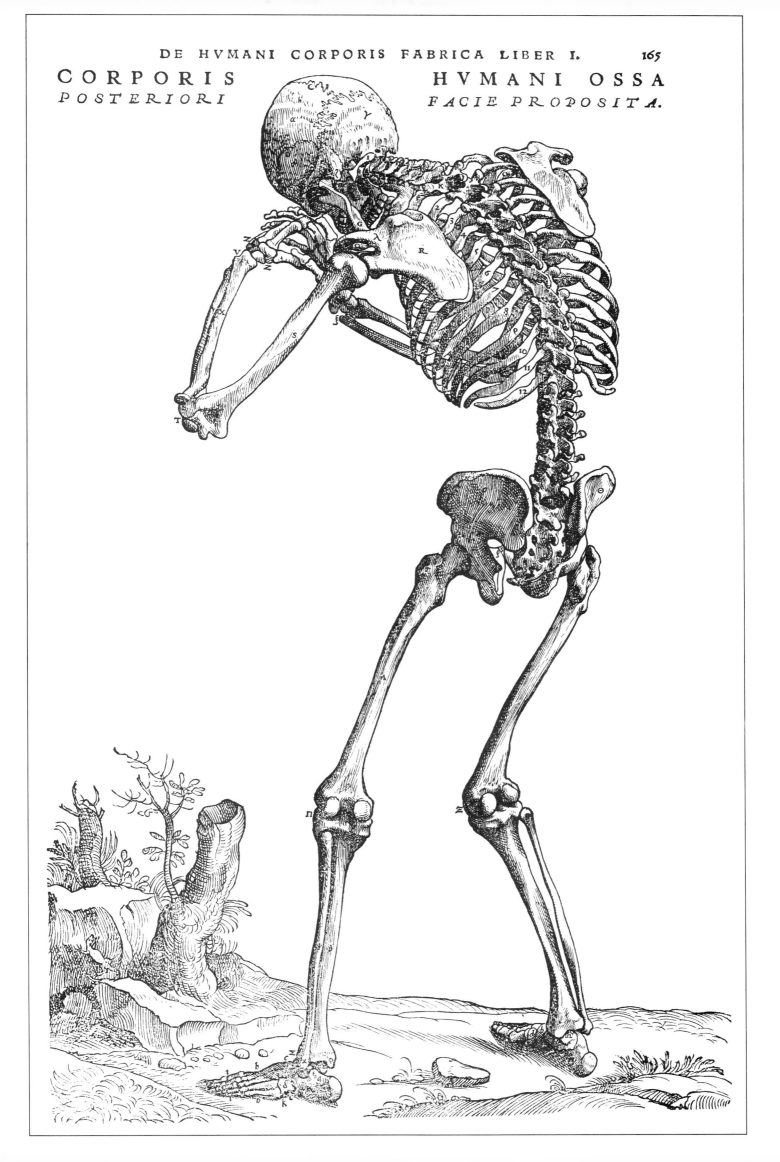

Anatomical Perspective: Albrecht Dürer

Albrecht Dürer (1471-1528) was intrigued with the structure of the human figure for most of his artistic life. His powerful engravings and drawings probed the human form for answers to his questions about the true nature of man. He spent a lifetime observing the design of man. In the end, he became at least as familiar with the human body as any physician of his day.

Born in Nuremberg, Germany, Dürer apprenticed to his father as a goldsmith before setting out on his own career as an artist. When he was twenty-three years old, he decided to travel to Italy where he could study with the "masters" of proportion and perspective who themselves had been drawn there by the inspiration of Renaissance inquiry.

Dürer was deeply influenced by the Renaissance school of naturalism in art. He believed that by careful observation the artist would move closer to understanding the true spirit of his subject. It was the expression of this true spirit that was the artist's task.

His artistic purpose, coupled with a genuine admiration for the human figure, fired Dürer's interest in the study of human anatomy. He dissected his first cadavers for the purpose of study while he was in Italy. He also snatched up some of the first copies of Euclid's scientific books off the printing press. If Dürer was going to represent a subject, he needed to know his subject's dimensions from every angle.

Later in his life, Dürer was to devote a good deal of his time to producing a book of his own anatomical perspectives of the human figure for artists. He also wrote several other scientific treatises which he unfortunately did not live to see in print.

If Leonardo was the "universal" Renaissance man of Italy, Dürer was surely his counterpart in the North. On the subject of nature as his inspiration, Dürer said, "For in truth art is implicit in nature, and whoever can extract it possesses it." Like Leonardo, he saw the art in nature, and wanted in turn to ensure that his art was true to its natural source.

Dürer's interest in human anatomy grew out of his quest for scientific knowl-

edge in general. This quest was for him inseparable from his purpose as an artist. There was exemplified in him that mysterious process by which pure observation can be transmuted into profound insight. Albrecht Dürer believed that this was the tradition out of which flowed both the inventions of science and the inspirations of art.

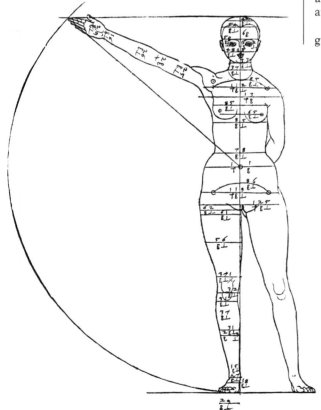

OPPOSITE: A detail from a study of a child's head, 1506, by Albrecht Dürer. Bibliothèque Nationale, Paris. LEFT: A plate from Dürer's book of anatomical proportions of the human body, *Alberti Dureri : clarissimi pictoris et geometrae de symetria partum . . .*, 1532. ABOVE: A drawing Dürer sent to his physician showing where he felt some pain, c. 1500. Kunsthalle, Bremen, Germany.

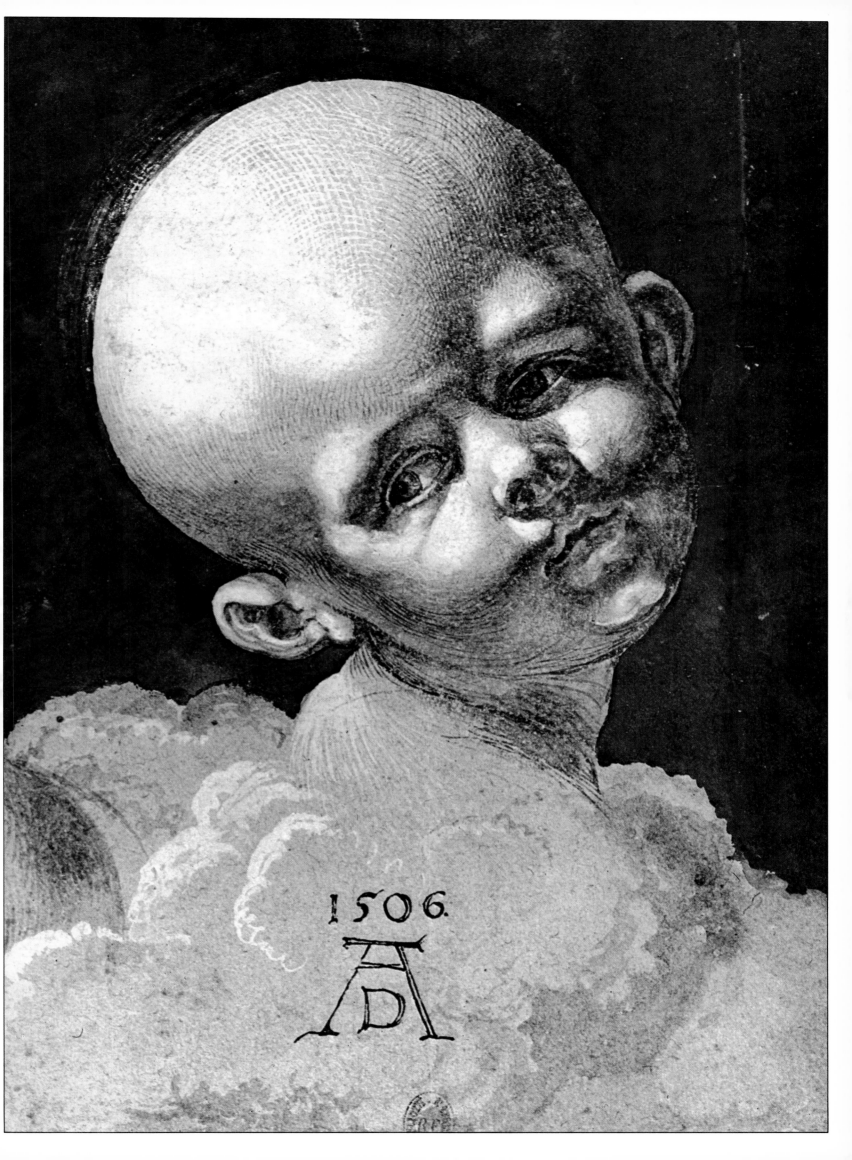

The Barber Surgeons of England

While the rest of Europe had already been swept by a tremendous surge of learning in the field of human anatomy, sixteenth-century England was just beginning to experience the repercussions of Renaissance inquiry. Vesalius' work *De Humani Corporis Fabrica* had reached England shortly after it was published in 1543, and by 1545 several pirated editions were already in circulation. In 1540, Henry VIII had granted the Barber Surgeons permission to dissect the cadavers of four criminals a year for the purpose of studying the human body. Although no major advances were to be made by this group, several Barber Surgeons produced illustrated texts describing the state of anatomy at that time.

A Barber Surgeon, Thomas Vicary, published the first printed book of anatomy in England around 1548. The Barber Surgeons were a group of barbers who had been licensed to practice surgery and dentistry. They were considered inferior practitioners to the members of the Royal College of Physicians. Yet it was the Barber Surgeons who were the first to be officially allowed to perform dissections. It would appear that the English still clung to the old notion that the work

THE ANATOMIST OVERTAKEN by the WATCH ... CARRYING OFF MISS W— in a HAMPER

of dissection was not the most desirable line of research or study. Not until 1565 did Queen Elizabeth decree that the Royal College should perform dissections on human cadavers.

John Bannister (1533-1610) published a work on anatomy while he was lecturer on the subject for the Barber Surgeons. Most of his work was copied from Vesalius. However, he also executed an ivory carving of a skeleton and a few sheet drawings which were among the first examples of an Englishman's independent research.

Despite the lack of original research from this period, England was to produce the great William Harvey, whose work on the heart in the seventeenth century was to transform the science of human anatomy all over the world.

OPPOSITE: John Banister delivering the visceral lecture at Barber-Surgeons Hall, London, 1581. *John Banister's Anatomical Tables* (Hunter Manuscript 364). Courtesy of the Glasgow University Library. ABOVE: *The Anatomist Overtaken by the Watch*, 1775, by Thomas Rowlandson. Courtesy of The Trustees of the British Museum. As this satire points out, illegal body snatching had reached drastic proportions by the eighteenth century. Despite punitive laws, human cadavers were in such short supply for the purpose of anatomical research, it was not unusual for anatomists themselves to take to creeping around graveyards in the dead of night, hoping to discover a freshly interred body to steal.

Anatomia scientiæ dux est
aditumque ad dei agnitio-
nem præbet.
Iohannes Banister Ætatis
suæ Anno 48

Anno Domini 1581

Tendit in ardua Virtus

De præscientia Dei

China: The Yin and the Yang

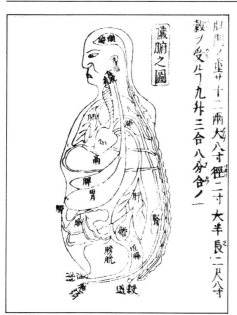

The Chinese school of human anatomy was rooted more firmly in philosophy and tradition than it was based upon investigation. From their earliest recorded history, the Chinese have opposed disfigurement of the body. Up until recent times, the practice of human dissection was strictly forbidden. Their only contact with the inner structure of the body had been through wounds and injuries.

The Chinese approached the structure of the human body the same way they approached the structure of all life. There was no spirit of exploration and discovery. There was no necessity to experiment. The functioning of the universe depended upon the balance of the principles of Yin and Yang. These principles existed in every living force. Therefore, the study of the human body had more to do with identifying Yin and Yang than with describing the body's structure.

In ancient China, the human body was said to be composed of five Zō and six Fu. The Zō consisted of the principal internal organs such as the heart, the liver and the lungs. These were referred to as the "storehouses." Each Zō corresponded to an element of the universe and ruled a season, a taste and a color. The Fu assisted the Zō and were called the "manor houses." The Fu were made up of organs such as the stomach, the large intestines and the gallbladder. Besides their other properties, the Zō and the Fu each housed a special animal. The white tiger was said to live in the lungs and the dragon in the liver.

The principles of Yin and Yang functioned as the circulatory system. There is no question that the Chinese had some notion of the relationship between the heart and the circulatory system at least one thousand years before the famous Dr. William Harvey did his experiments on the heart in England. The legendary Yellow Emperor Huang Ti (2698-2598 B.C.) is credited with mentioning the Chinese theory of the circulatory system in the manuscript *Nei Ching*, which he is said to have written with the aid of his physician and court scholar. Evidently, the Chinese believed in a theory of double circulation. That is to say, there was a system which consisted of the "spirits," the carriers of the Yin, and the blood, the conveyor of the Yang. This system distributed itself throughout the body, beginning at 3:00 A.M. (the hour of the tiger) in the lungs. It took twenty-four hours to complete its whole cycle of the body.

The ancient practice of acupuncture was established to maintain the balance of the Yin and the Yang. Acupuncture was said to release all the bad secretions in the body and get rid of any tissue obstructions. There were three hundred sixty-five vital points in the body, each of which corresponded to a day of the year.

Needles were inserted into the various vital points to restore the balance of the Yin and the Yang. The Chinese continue to practice acupuncture today.

The only other aspect of the body's structure upon which the Chinese placed emphasis was the areas of the pulse. According to the Chinese, there were hundreds of pulses throughout the body. For example, each Zō and each Fu had a pulse. There were at least thirty different pulse areas on the arm alone. One could read the entire well-being of the body by taking the different pulses. Although there has never been any evidence that these pulses do in fact exist, Western physicians do not dismiss the theory lightly. There have been too many cases in which a disorder was correctly diagnosed by the Chinese reading the body's pulses.

The ancient school of human anatomy survived in China almost wholly intact until just recently. China's reverence for tradition maintained this approach to human anatomy even in the face of Western research and the dawning of new knowledge about the structure of man.

OPPOSITE: A lithograph by Langlume, from an unknown Chinese source, of a tsoë-bosi, a model which contains all the points where acupuncture and moxas should be applied, early nineteenth century. The New York Academy of Medicine. The name tsoë-bosi means priest-figure and is derived from the fact that the head of the model is shorn like that of an oriental priest. ABOVE, LEFT: A classical Chinese anatomical figure representing the five Zō and six Fu, 1685. From the book *Shinkyū bassúy*. Courtesy of Gordon Mestler. LEFT: A bronze model used for teaching the skill of moxabustion, 1027. Courtesy of Gordon Mestler. Moxabustion is the technique of placing burning tapers (moxas) dipped in plant oils against the skin at certain crucial points along the body to relieve pain.

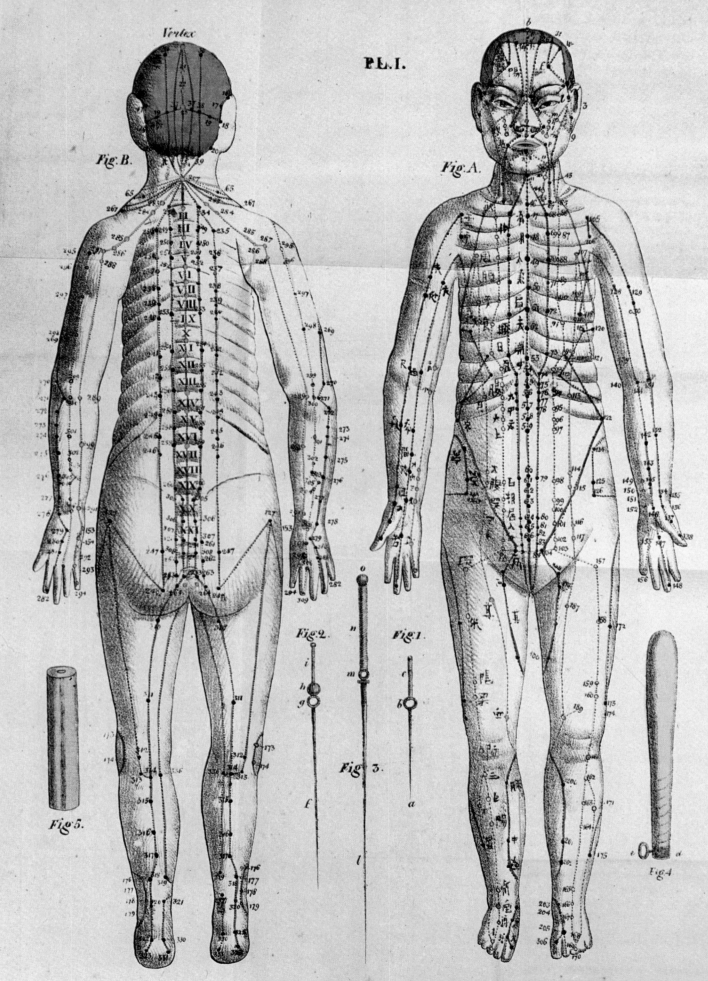

PL. I.

Fig. B.

Vertex

Fig. A.

Fig. 2.

Fig. 1.

Fig. 3.

Fig. 5.

Fig. 4.

Lith. de Langlumé

Tsoë-Bosi.

Japan: The Kaitai Shinsho

The origins of human anatomy in Japan were based largely upon Chinese and Dutch influences. The Japanese established no authentic system of their own. So many aspects of Japanese culture had already been adopted from the Chinese, it seemed natural that they would rely upon the Chinese for their understanding of human anatomy. The Chinese model remained the standard for the study of the human figure until well after the Dutch arrived with their anatomical books. These Western books contained new material which strongly conflicted with the older Chinese version of anatomy. Not until a Japanese physician named Genpaku Sugita took it upon himself in 1771 to make a comparison between the two versions, were there any new developments in human anatomy in Japan.

The earliest records of anatomical knowledge in Japan date at least as far back as the sixth century. It was the habit at that time in Japan to send Buddhist monks to China to study with Chinese scholars. Upon their return, they were expected to teach the Japanese what the Chinese had taught them. When these monks brought back the Chinese version of human anatomy, the Japanese were only too ready to absorb the teaching just as it was presented to them. It was not their tradition to question the information; it was only for them to learn it.

By the eighteenth century, there had been enough contact with outsiders from the West, especially the Dutch, for the Japanese to have become acquainted with Western versions of anatomy. Although the Japanese government had banned all contact with Westerners in 1603 because it feared Christian influences upon Japanese society, it was unable to prevent the circulation of various Western medical dictionaries and reference books. The anatomy contained in these books thoroughly disputed the ancient Chinese models of the structure of the human body. The Japanese physicians became confused. Their government and history supported the standard model of the Chinese. But this new information was beginning to make the medical men of Japan uncomfortable.

A Japanese physician, Genpaku Sugita, decided in 1771 that he wanted to find out which version of human anatomy was correct. He had obtained a copy of the Dutch translation of the German J. A. Kulmus' *Anatomische Tabellen* (Anatomical Tablets). He had been appalled at the discrepancies in the internal structures of the body. He felt that he owed it to his patients to discover the truth.

Upon being invited to attend a dissection, Sugita invited two of his colleagues to come with him and compare the Dutch anatomical model with their standard Chinese model. The doctors were devastated by the results. The Chinese model upon which they had depended for all these hundreds of years was clearly in error. Sugita described in his journal how upset they had been at having been so passive and having waited until then to question what they had learned.

After their experience with the dissection, Sugita and his friends decided they would take whatever amount of time was necessary and translate the Dutch anatomical atlas for the benefit of the rest of the physicians in Japan. It was no easy task, as none of them spoke a word of Dutch. They spent three years on the project—but their work profoundly affected the development of human anatomy in Japan.

The book they put together came out in 1774 and was called the *Kaitai Shinsho* (A New Book of Anatomy). Its five volumes

contained the translation of *Anatomische Tabellen*, and was illustrated with curious adaptations of the Dutch anatomical images. The drawings were exact copies as far as the structure of the body was concerned. However, to make the figures appear more Japanese, they had drawn in Japanese heads and eyes on top of the heavyset Western bodies. Whatever the curiosities of the book, its publication marked the beginning of the modern era, and the opening of Japan to the advances in anatomical knowledge of the West.

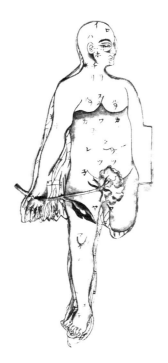

OPPOSITE: A Japanese woodblock print of the digestive system, early nineteenth century, artist unknown. LEFT: A Japanese drawing after a Dutch original, c. 1774. From *The Dutch Atlas of Anatomy of the Whole Body*. ABOVE: A Japanese interpretation of a Western title page in the *Kaitai Shinsho*, 1826. All courtesy of Gordon Mestler.

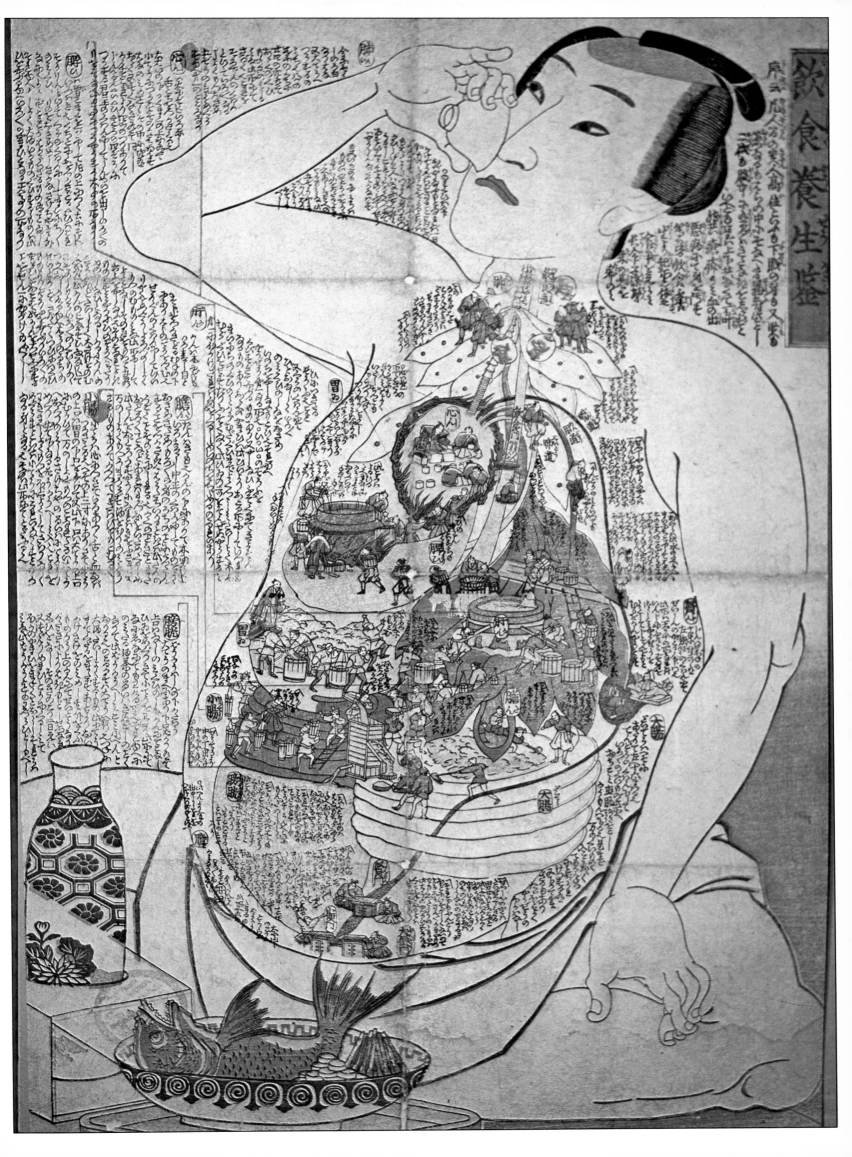

Anatomical Theater: Rembrandt van Rijn

"Here addresses us the eloquence of learned Tulpius, while with nimble hands he dissects livid limbs. Listener, learn for thyself, and as thou turnest from one part to another, believe that even in the smallest particle God is enshrined."   Caspar Barlaeus, poet, 1639.

In Europe, the beginnings of what can only be described as anatomical theater can be traced back to the end of the fifteenth century, when the first public anatomies were held. The early public anatomies were simple affairs which were held in the town square and attended by whoever happened by. As the years went on, what had started out as a simple affair for the benefit of the public turned into a dramatic performance for a paid audience. Public anatomies developed into large-scale events which had been planned months in advance of their date and were freely advertised in the handbills of the day. By the seventeenth century, elaborate amphitheaters were being constructed for the purpose of these performances. The anatomists, who were elegantly robed, in

time achieved reputations not unlike those of well-known performers of today.

The occasion of a public anatomy created tremendous excitement and attracted huge crowds. Tickets to the event were in great demand, despite their extravagant cost. The public anatomy became an annual occasion, marking the beginning of winter. It was important that the dissection of the cadaver take place in cold weather because of the perishable nature of the body. There were usually three stages to the event, with festivities planned over a four- or five-day period. The first stage occurred the day the criminal to be dissected was executed. The second stage consisted of the dissection of the cadaver, and the third involved a torchlight parade and a semiprivate banquet for members of the physicians' guild. The dissection itself was usually performed at sunset and took three to four days to complete.

In 1632, when Rembrandt painted *Professor Tulp's Anatomy Lesson,* anatomical theater was at its height. Nicholas Tulp was a highly respected anatomist of his time. Rembrandt himself had probably witnessed many of his performances, as he frequently attended the public anatomies of the period. His painting of Nicholas Tulp's lecture evoked the dramatic quality of public anatomies, and it confronted the relationship between science and art, which was preoccupying the minds of many learned men of that period.

The public's enthusiasm for the public anatomy grew out of its theatricality. The performance of the anatomist was like that of the actor upon a stage. His appearance, his delivery, his style were all ingredients of his popularity. And then there was the material itself. The impact of observing the intricate structure of the human body could not easily be rivaled for its dramatic appeal.

OPPOSITE: *Professor Tulp's Anatomy Lesson,* 1632, by Rembrandt van Rijn. Mauritshuis Museum, The Hague. LEFT: The dissection room at Leyden University, 1616, by Petrus Paaw. The Metropolitan Museum of Art, Gift of Lincoln Kirstein, 1953. ABOVE: The anatomical theater at Sweden's Uppsala University, built by the famed anatomist Olof Rudbeck in 1662. Photograph by David Balderston.

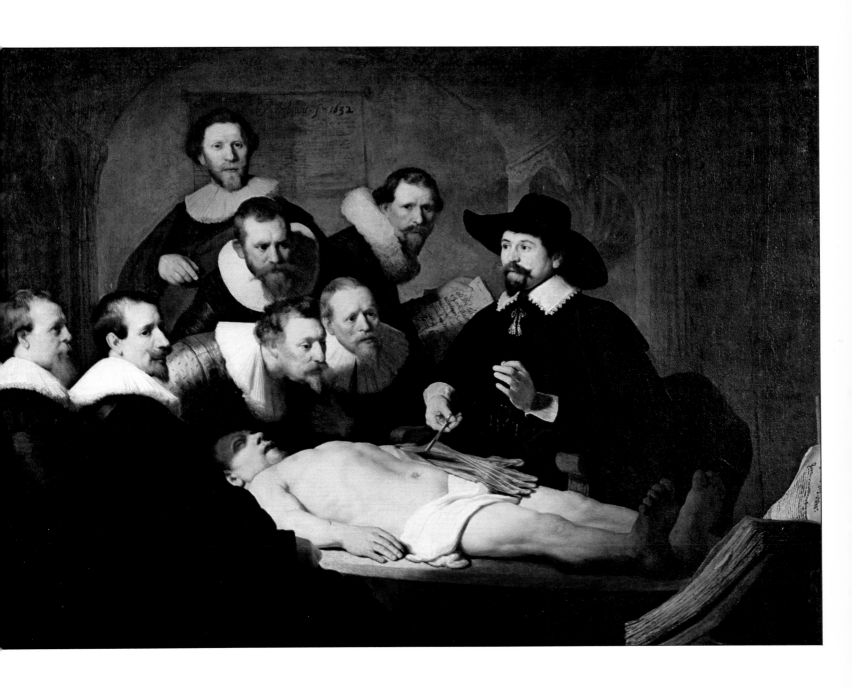

Engraved Images

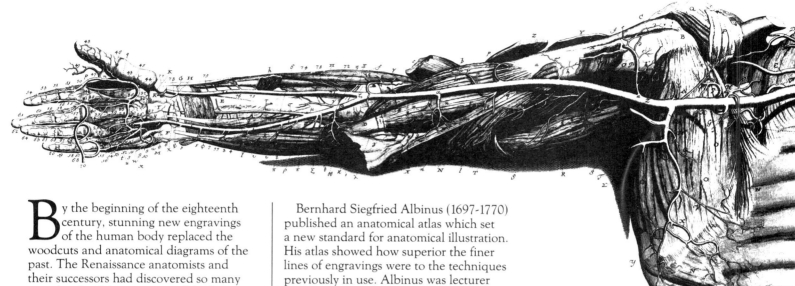

By the beginning of the eighteenth century, stunning new engravings of the human body replaced the woodcuts and anatomical diagrams of the past. The Renaissance anatomists and their successors had discovered so many new details of the structure of the human body that woodcuts and the other previous illustrative techniques could not produce fine enough lines to articulate these details. Engravings, on the other hand, were made up of a series of tiny lines which were capable of defining the minutest detail. There arose at this time a new pride in achieving artistically perfect engravings which also were scientifically accurate representations of the structure of the human body.

Bernhard Siegfried Albinus (1697-1770) published an anatomical atlas which set a new standard for anatomical illustration. His atlas showed how superior the finer lines of engravings were to the techniques previously in use. Albinus was lecturer and then professor of anatomy at Leyden University in Holland. He chose one of the finest engravers of that period, Jan Wandelaar, to execute his anatomical studies. His drawings were done from actual dissections and then handed over to Wandelaar to engrave on plates.

Albinus spared no amount of time or energy on perfecting his drawings. He was particularly interested in comparing the proportions of human skeletons with the rest of the living world. Sometimes he would set his figures against animals, sometimes trees, and occasionally he would use stone walls or pieces of sculpture. His work with bones and with the female reproductive system was especially good. The result of all his efforts was an atlas on the structure of the human body which has been called the finest example of artistic workmanship and science to come out of that period.

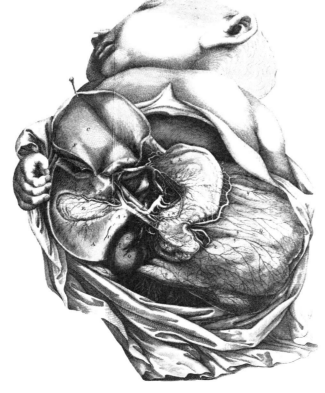

OPPOSITE: An engraved plate from *Uteri Gravidi*, 1753, by Bernhard Siegfried Albinus. Yale Medical Library. LEFT: An engraving of an infant, 1752, from *Icones Anatomicae*, by Albrecht von Haller. Yale Medical Library. ABOVE: Von **Haller** engraving of an arm from the same collection.

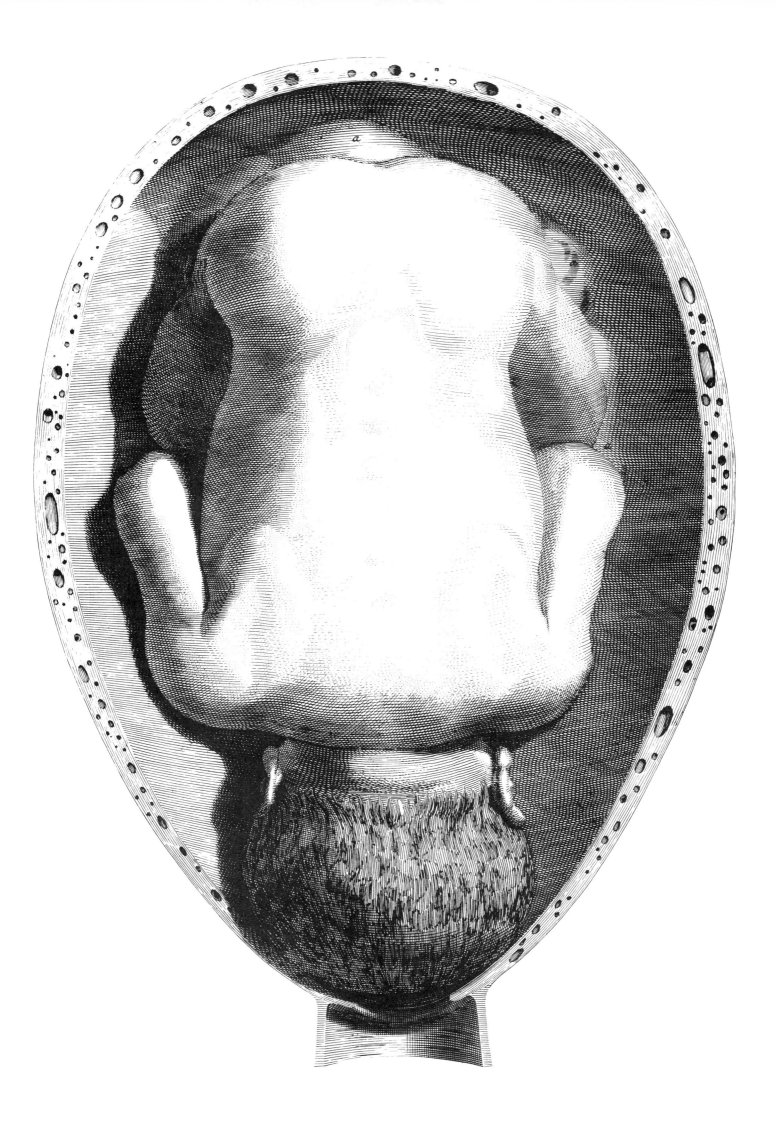

The Arabic Contribution

The Arabic-speaking world played an important role in the history of human anatomy. Throughout the Middle Ages, Arab armies crossed the desert and conquered large sections of the Middle East. As a part of the spoils of war, the Arab invaders seized most of the anatomical observations of Western civilization which had been recorded and stored in the great libraries of Alexandria, Egypt. The Arabs carried this material back to their own states and proceeded to translate it from Greek into the Arabic language.

When the Arabs translated these anatomical records, they preserved a tradition of Western scholarship which the Christian Church was, at that time, making every attempt to wipe out. Scholarship in the pursuit of worldly knowledge was not acceptable under Christian dogma. In particular, the study of the human body was heretical for its emphasis on man as opposed to the Church. The Church actually forbade the study of human anatomy, destroying any anatomical material that came to its attention. Had it not been for the Arab invaders, the anatomical observations of early

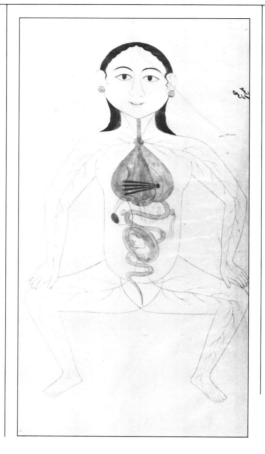

Western civilization might never have survived intact. It wasn't until the thirteenth century that the Arabic translations found their way back to Europe and, in turn, were translated into Latin.

The Arabs, as it happened, benefited tremendously from their seizure of the Western anatomical documents. They had never developed a system of their own because their religion forbade pictorial representations of the human body. The practice of human dissection was also strictly prohibited. There was not even a course in anatomy in the curriculum of medical schools. Anatomical references to the human body were limited to verbal descriptions of its parts and the sites of various organs. The diagrams and other anatomical images that have been referred to as Arabic were, more than likely, copies or illustrations from the Western records. The Arabs, however, were keenly interested in learning about human anatomy and eagerly devoured the Western world's observations on the structure of man.

OPPOSITE: A schematic diagram of the eye from an Arabic manuscript, c. 1714. Bibliothèque Nationale, Paris. ABOVE: The arterial system of a pregnant woman from a Persian manuscript, 1596. MS. Fraser 201, fol. 1013 r. Department of Oriental Books, Bodleian Library.

وهذه صورة العين وطبقاتها ورطوباتها واعصابها وعضلاتها والتقاطع

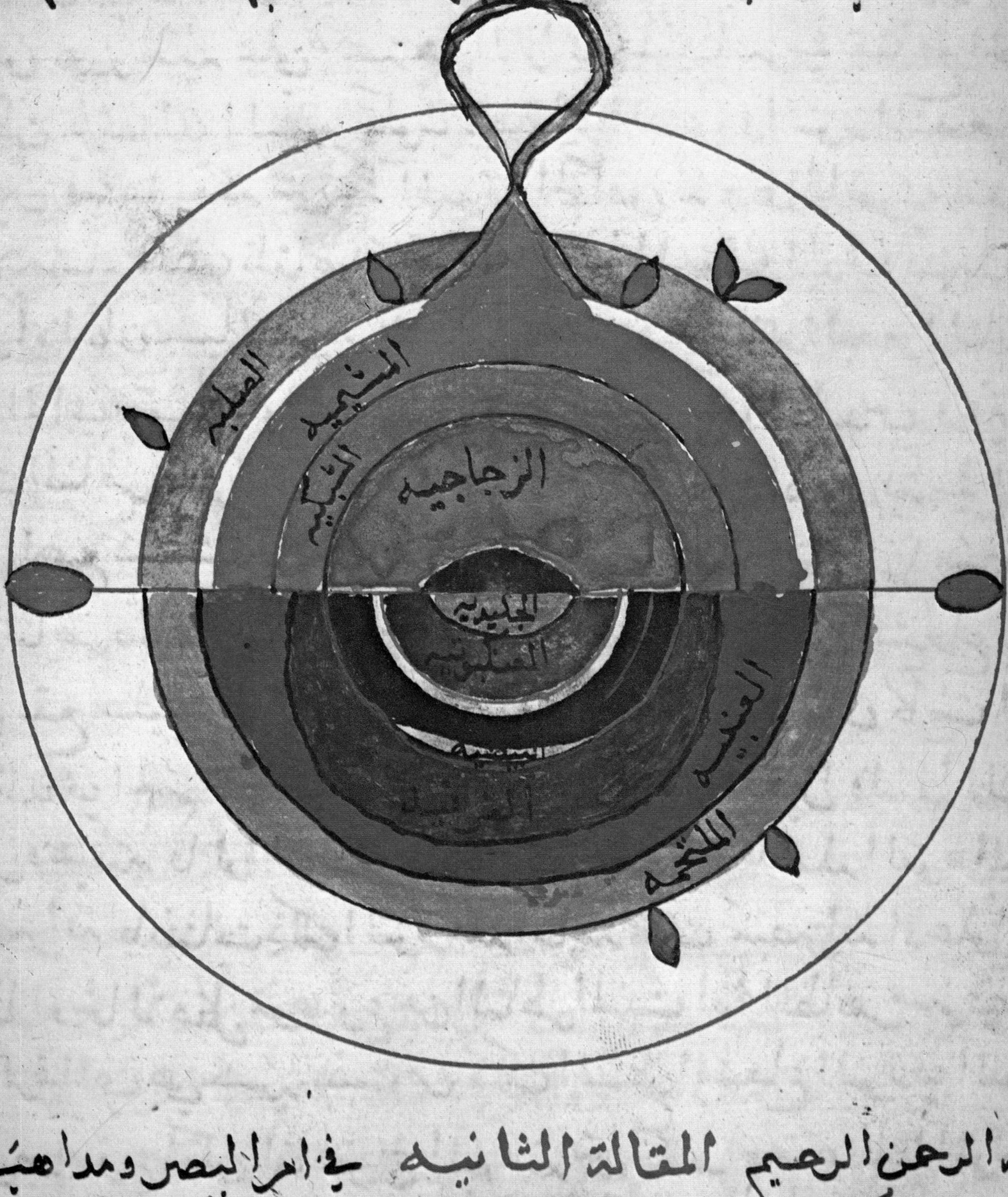

الزجاجيه

المشيميه

البيضيه

العنكبوتيه

الصلبه

الملتحمه

بسم الله الرحمن الرحيم المقالة الثانيه في امر النص ومذاهب الحكما في كيفية ادراك المبصرات وهي خمسة ابواب الباب الاول اعلم ان مذاهب الحكما في كيفية ادراك المبصرات وهي ثلاث مذاهب المذهب الاول مذهب الرياضيين وهم القائلون بخروج الشعاع من العين والمذهب الثاني مذهب من يرى بتكيف الهوى الخارج والمذهب الثالث مذهب

Anatomical Colorist

The Frenchman Gautier D'Agoty (1717-1786) produced some of the most beautiful color studies of human anatomy ever to have been done. He was also the first anatomical illustrator in history to use color extensively in his work. Unfortunately, the technique of mezzotint which he used was not precise enough to render the most intricate details of the structure of the human body. D'Agoty's work was greatly admired for its striking use of color, but it fell short of being taken seriously by the scientific world.

D'Agoty's studies of pregnancy were among the very few that seem to express both the mystery and the dignity of the female experience. Historically, the anatomy of women had been limited to the example of the pregnant model. The most ancient diagrams had pictured her squatting, with no reference to her external female organs. Through the years, she might have gained a few more external details, but the scientific community's attitude remained the same. Woman's anatomy was not important except in the context of pregnancy, and even then only from a comfortable distance.

D'Agoty, on the other hand, was perhaps slightly less precise, but his recognition of the experience of creating another human being within the human structure imbued his work with a special quality. No matter what period of pregnancy he was depicting, he somehow managed to render the structure of the female figure as if it were the subject of an artist's dream. He certainly had not meant to sacrifice accuracy for effect, and in fact, his work fell just short of the standard. But D'Agoty's technique aside, his anatomical studies of pregnancy are remarkable for their appreciation of the female experience.

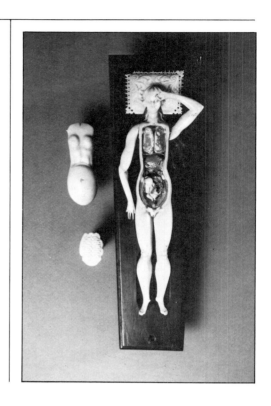

OPPOSITE: The muscle system of a pregnant woman, 1773, by Jacques Fabian Gautier D'Agoty in *Anatomie des Parties de la generation de l'Homme et de la Femme.* Yale Medical Library. ABOVE, LEFT: A satire of a young medical student with his girl-friend and a "skeleton in the closet," mid-nineteenth century, by Paul Gavarni. Bibliothèque Nationale, Paris. ABOVE: A German anatomical ivory mannequin of a woman with child, seventeenth century. The New York Academy of Medicine, Bequest of Dr. Jerome Pierce Webster, 1975.

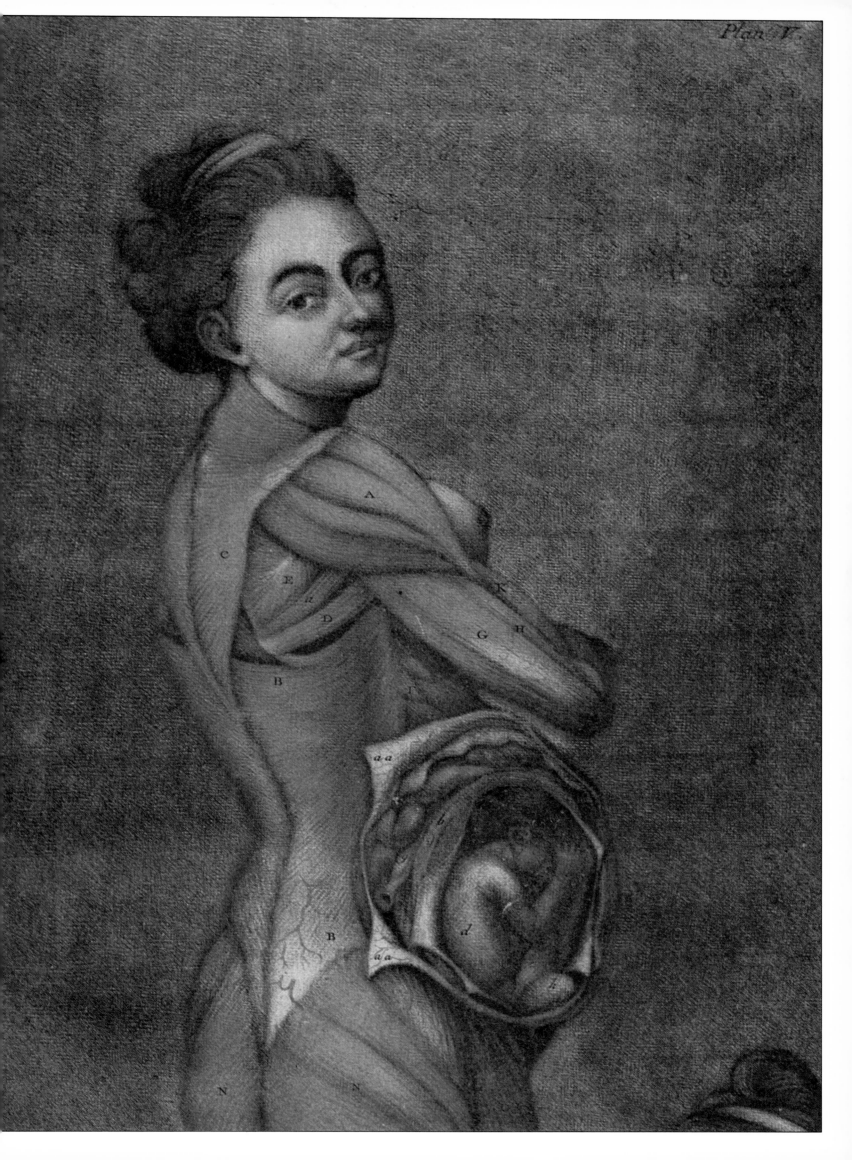

Plan. V.

The British Anatomists

Although late in embracing European advances in Renaissance anatomy, England nevertheless produced a share of dedicated anatomists whose work did affect the international course of human anatomy. The most famous Englishman in the study of anatomy was the great Dr. William Harvey (1578-1657). His research on the heart and the circulation revolutionized all prior concepts as to their relationship and functions. He also thoroughly altered the method of studying the structure of the human body. He chose to isolate each organ and observe its structure and function independently of the body as a whole. For the first time, anatomists began to consider the relationship of the various parts of the body as well as their structure. William Harvey's method is the same one adopted by anatomists of today.

Another important English anatomical study was contained in an atlas that John Bell published in 1794. His engravings of the parts of the human body achieved a new level of accuracy and artistic perfection. The number of studies done on each bodily structure as well as the different angles of the studies made this a particularly thorough reference for the study of human anatomy. This atlas was also perhaps among the last of the premicroscopic reference works. After this period, the microscope would reveal a view of the human structure that the naked eye could not see. Beneath the surface of man, there was yet another whole structural world to observe and attempt to understand.

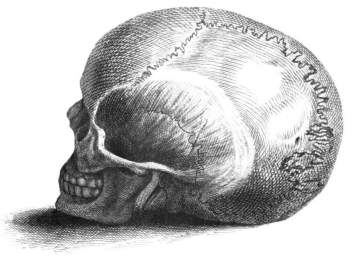

OPPOSITE: An engraved plate from *Anatomy of the Bones, Muscles and Joints* by John Bell, 1794, LEFT: A detail from a plate in *Anatomy of the Bones, Muscles and Joints*. ABOVE: A plate from *Thesaurus Anatomicus* by Frederic Ruysch, 1701. The curious composition of this anatomical plate was a favorite decorative scheme of Dutch anatomists in the seventeenth and eighteenth centuries. All from The Metropolitan Museum of Art, Gift of Lincoln Kirstein, 1953.

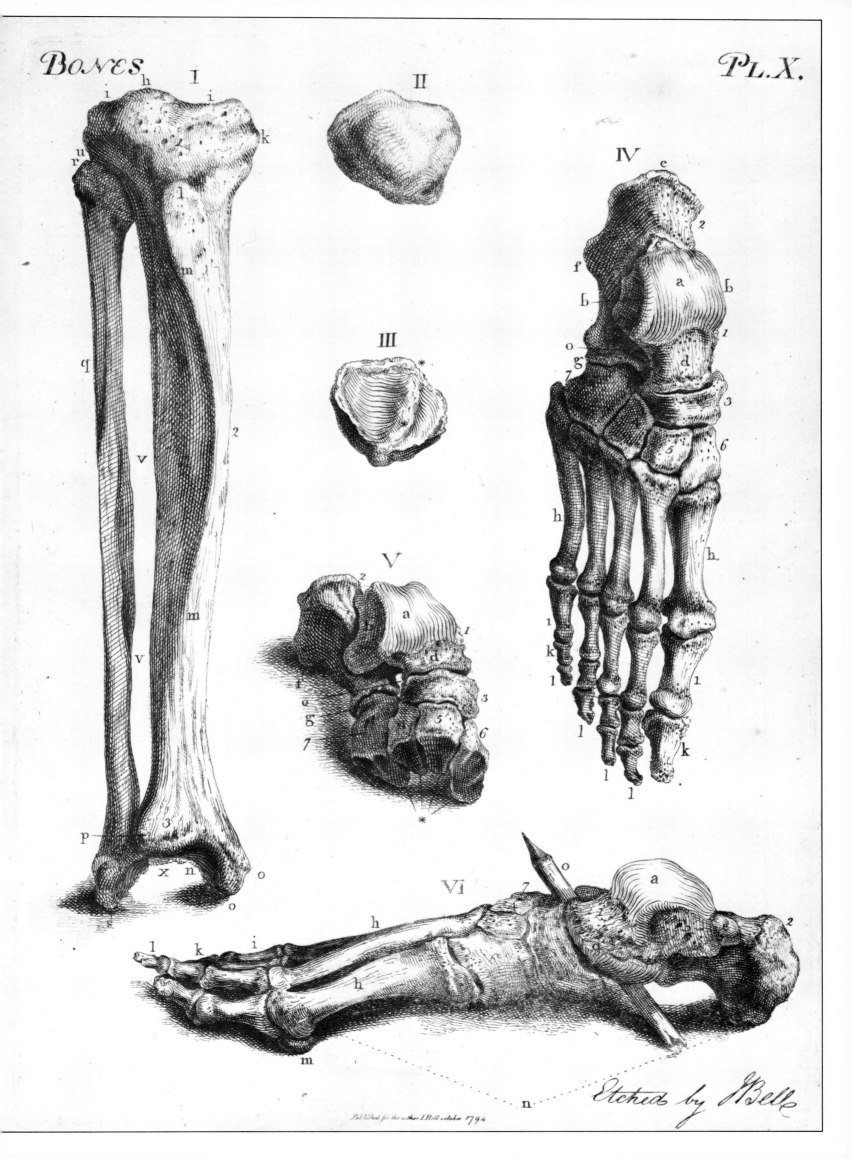

BONES

PL.X.

Etched by J. Bell

Published for the author J. Bell october 1794

Artistic Anatomies

In the first half of the sixteenth century, illustrated anatomies produced by artists were made available to the public. The first ones were published on what were known as "fugitive sheets." The earliest "fugitive sheets" were actually done by anatomists for university students who were just beginning their study of the human body. These individual illustrations were not as detailed or thorough as those in the textbooks of the day and were considered to be inferior studies. However, when the first artistic "fugitive sheets" appeared, quite a change occurred. In place of the loose university illustrations, the artists turned out highly detailed, beautifully rendered studies of the human body. These artistic "fugitive sheets" were so superior that the universities frequently circulated them in lieu of their own.

Over the years, many major artistic anatomies have been published. One such study was undertaken in the nineteenth century by a doctor turned artist named William Rimmer. In 1877, Rimmer published *Art Anatomy*, a collection of over nine hundred drawings of the human form and its various parts.

William Rimmer is thought to have been born in the mid 1800s in the Boston,

THE ANATOMIST.

Massachusetts, area. He had always been interested in art, but he had decided he could earn a better living to support his wife and children as a doctor. Unfortunately, since most of his efforts were devoted to treating the poor, he never earned a cent. At his friends' urging, he gave up his medical practice, took a job as a lecturer on anatomy for an art school and undertook to devote his full time to art.

Rimmer has been described by his contemporaries as one of the finest anatomy lecturers of his day, and his classes are said to have been well attended. However, as much as he admired the human form, Rimmer evidently held the human body at a certain distance in his classes. He could not cope with the concept of the nude model, particularly where his female students were concerned. His inability to confront the nude model

in his classes, however, certainly did not affect the quality of the anatomical studies in his book. Where he failed as a lecturer, he more than succeeded as a draftsman. He rendered the structure of man, woman and child in intricate detail and from the most intimate angles. William Rimmer's *Art Anatomy* is by far one of the most thorough surveys of the human form to have been made by an artist in modern times.

OPPOSITE: Anatomical studies of a young child, 1877, in *Art Anatomy*, by William Rimmer. LEFT: *The Anatomist*, a satire after Thomas Rowlandson, 1811. Courtesy of The Trustees of the British Museum. ABOVE: *Anshutz on Anatomy*, 1912, by John Sloan. Collection Whitney Museum of American Art, New York.

Infant Proportions

Stages of Development.

Nos 257.258. In comparison with the proportions of an
adult, the head of an infant is large in proportion to
the size of the body; the body large in proportion to the
size of the limbs; limbs large in proportion to the size of the
hands and feet. In the first stage of development the
spinal muscles are too feeble in their action to maintain the
body in an erect position. The back convex, the abdomen
large and bag-like, and the legs and arms bent outward.
Suppositional= The expression serious and void of passion.
The mind curious and attentive.

Nº 257.

1st Stage.
Plan.

Nº 258.

2d Stage.
Plan.

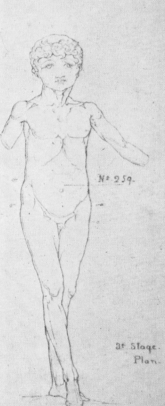

Nº 259.

3d Stage.
Plan.

Nº 260.

2d Stage=

Artist and Anatomist: Thomas Eakins

Thomas Eakins (1844-1916) believed that the study of anatomy was essential to his work as an artist. Like Leonardo, he put his trust in the real rather than the idealized. His observations were the material of his creations. The magic, the spirit, the vitality were all to be found in the pure form of a subject. For Thomas Eakins, it was as if science and art were merely expressions of each other.

Eakins' lifelong commitment to the study of human anatomy began when he was sixteen years old. Having graduated from his Philadelphia high school, he attended anatomy classes at Jefferson Medical College while studying art at the Pennsylvania Academy of the Fine Arts. At Jefferson he did his first dissections of human cadavers, patiently observing how the whole mysterious human figure fitted together. For a short time, he even seriously considered becoming a surgeon instead of a painter. In the end, he decided not to go into medicine, and in 1866 he sailed for Paris to further his art studies.

While in Paris, Eakins pursued his interest in anatomy by practicing dissection on human cadavers at the Ecole des Beaux Arts. Several years later, he returned to Philadelphia and began once again to attend the anatomy lectures at Jefferson. Many of the subjects he chose to paint during these years came out of his experiences at Jefferson.

In 1876, when Eakins himself began to teach at the Pennsylvania Academy of the Fine Arts, he required his students in his Life Drawing class to attend a course on human anatomy. He wanted them to actually take part in a dissection. As unpleasant as it was, and Eakins freely admitted there was nothing pleasant about handling a human cadaver, he was adamant in maintaining that there was no way to learn to draw the human figure without a firsthand knowledge of its structure.

Eakins became so well acquainted with anatomy that he did some original research of his own on muscles and joints. He gave a paper before the Philadelphia Academy of Sciences in which he presented his findings.

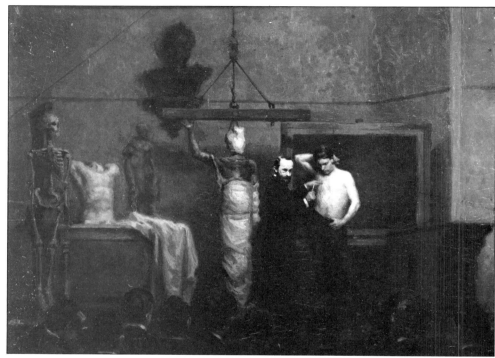

Walt Whitman, whose portrait Eakins painted, said of him once, "I never knew but one artist, and that's Tom Eakins, who could resist the temptation to see what they think ought to be rather than what is." This quality which so pleased Whitman was a problem with many of the other subjects who asked to sit for Eakins. Most of these subjects did not want, nor in many cases were they prepared, to accept themselves exactly as they appeared. But Tom Eakins was committed, despite the rash of criticism that followed him throughout his life, to reproducing what he observed to be the human form in its purest sense.

Eakins' dedication to the study of human anatomy brought about his dismissal from his prestigious teaching position at the Academy. One day in his Life Drawing class, Eakins removed the loincloth from one of his male models so that the class could accurately render the whole figure, as opposed to just the upper torso and the legs. Word of his action quickly spread around the Academy, and it was almost no time before the Academy fired him for such an improper and unnecessary act of indecency. So appalled was Philadelphia society by Eakins' exposure of the model that the Philadelphia papers carried editorials criticizing his lack of judgment and uncouth behavior.

Eakins was greatly moved by the beauty of the human body in its purest form, and his study of human anatomy was infused with tremendous enthusiasm. It was his appreciation of the natural beauty of the human body that was the source of his struggles against the artistic and social conventions of his day. Finally, this appreciation was also the source of a brilliant body of work which has been recognized since his death in 1916 as some of the finest art ever to come out of this country.

OPPOSITE: *Professor Gross (Gross Clinic)*, 1875, by Thomas Eakins. Courtesy of Jefferson Medical College, Philadeliphia. ABOVE: *Anatomical Lecture by Dr. Keen*, 1879, by Charles H. Stephens. Courtesy of the Pennsylvania Academy of the Fine Arts.

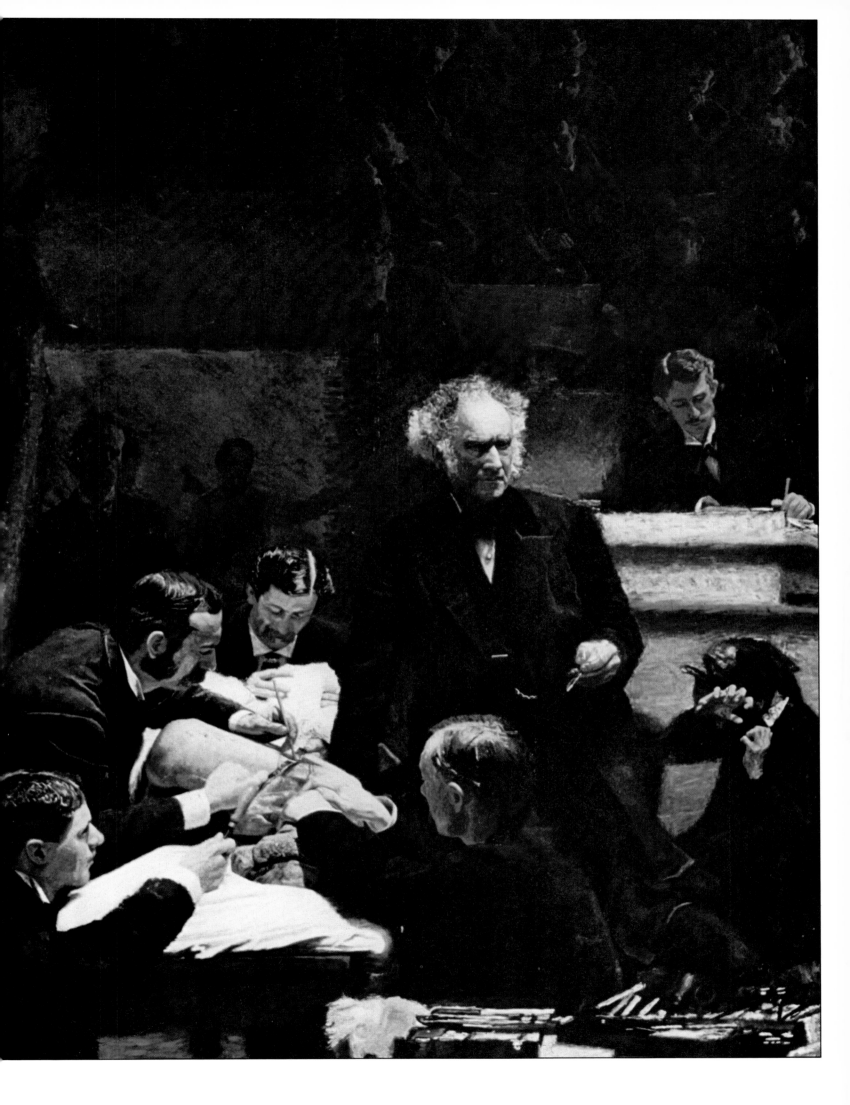

Microscopic Anatomy

"Where the telescope ends, the microscope begins. Which of the two has the grander view?" Victor Hugo, Les Misérables

The invention of the microscope revolutionized the science of human anatomy. Over the centuries, the study of human anatomy had made vast progress, going from the ancients' schematic images of man's external form to the Renaissance anatomists' investigations of the inner structure of the human body. Until this time, however, anatomy had meant gross anatomy—that is, observation of the body and its parts based upon its dissection. The microscope ushered in a whole new era of investigation and study. It revealed the presence of a vital new world of abstract shapes and structures that existed beneath the surface of the human body. The study of microscopic anatomy was the study of the human cell. It was to lead, in a short time, to the furthermost reaches of the cell and the study of molecular anatomy.

The first microscope was invented about 1603 by a Dutch spectacle maker named Zacharias Janssen. He constructed a telescope which could be rearranged into a microscope. The device was exhibited in 1608 at the Frankfurt fair in Germany and was bought by a friend of Galileo's as a present for him. Galileo wrote about the telescope/microscope when for the first time he aimed its long, tubular body at the heavens. The first real scientific work with the microscope, however, wasn't done until 1683, when another Dutchman named Antonie van Leeuwenhoek observed tiny spermatozoa swimming around on one of his slides. His lens had only very low magnification power, but it was enough to permit him to observe, in addition to the spermatozoa, the striations of muscles, as well as some bacteria from the mouth.

The early microscopes were light microscopes. The eye requires light to see; the microscope used lenses to magnify what was too small for the eye to see. The naked eye sees by assembling the light waves it receives into images. What the microscope does is take a beam of light, reflect it off a mirror, send it through a series of condensing lenses, and pass it through the image being observed. The image is then magnified by still another

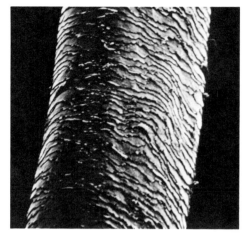

set of lenses which are focused into an eyepiece. The earliest microscopes could magnify an image two hundred or three hundred times at most. Today's microscope has the power to magnify images twenty-five hundred times. This means the human eye can see a detail that is about 1/125,000 of an inch across.

Exploring the universe of the cell, anatomists were at first startled by incredible new sights. Through the microscope's lenses, they observed the strange colors and weird shapes of a living world totally invisible to the naked eye. Over the years, the mysteries of the human cell have fascinated the whole of medical science. Even today, while much more is known about the cell, it continues to be a main focus for scientific research into the nature of human life.

The study of cellular anatomy posed as many questions as it revealed answers. Anatomists, probing the structure of the cell, began to look for new ways of delving even deeper into man. Light microscopes were improved through the years, but it was the invention of the electron microscope in 1932 that seriously advanced their research.

The first electron microscope was built in Berlin by Max Knoll and Ernst Ruska. Instead of using light beams, which have only a limited capacity to define images, Knoll and Ruska substituted electron beams, whose capacity is far greater. The first electron microscope worked in the same manner as the light microscope in that the electron beam had to shine through the image being observed. Instead of glass lenses, however, the beam passed through magnetic coils.

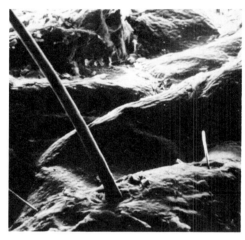

It was referred to as a transmission electron microscope.

From this first transmission electron microscope, Knoll moved on to the scanning electron microscope. In 1938, he came up with the idea that the electron beam could scan an image in such a way as to cause the image to release its own set of electrons, which could be picked up by a sensitive detector. The detector would then transmit the electron signal from the image onto a viewing screen. The system worked on the same principle as closed-circuit television. The scanning electron microscope has been refined over time, but the principle has remained the same.

The eye of the electron microscope has penetrated deep into man's innermost space to reveal the existence of the minuscule structure of the molecule. The study of molecular anatomy, in recent years, has been the main focus of anatomists. The observations being made today in molecular anatomy are easily as revolutionary as those of Leonardo and Vesalius at the time of the Renaissance. In both instances, ingenuity and inspiration were the ingredients that went into the creation of a most revealing portrait of the human structure and the nature of life itself.

OPPOSITE: A selection of different cells as the eye of the microscope reveals them, late 1970s. Courtesy of the Yale Medical School. ABOVE, LEFT: A strand of hair magnified under an electron microscope. ABOVE: The same electron microscope scans the variations in a single layer of skin. Woodfin Camp/Patrick Thurston photos.

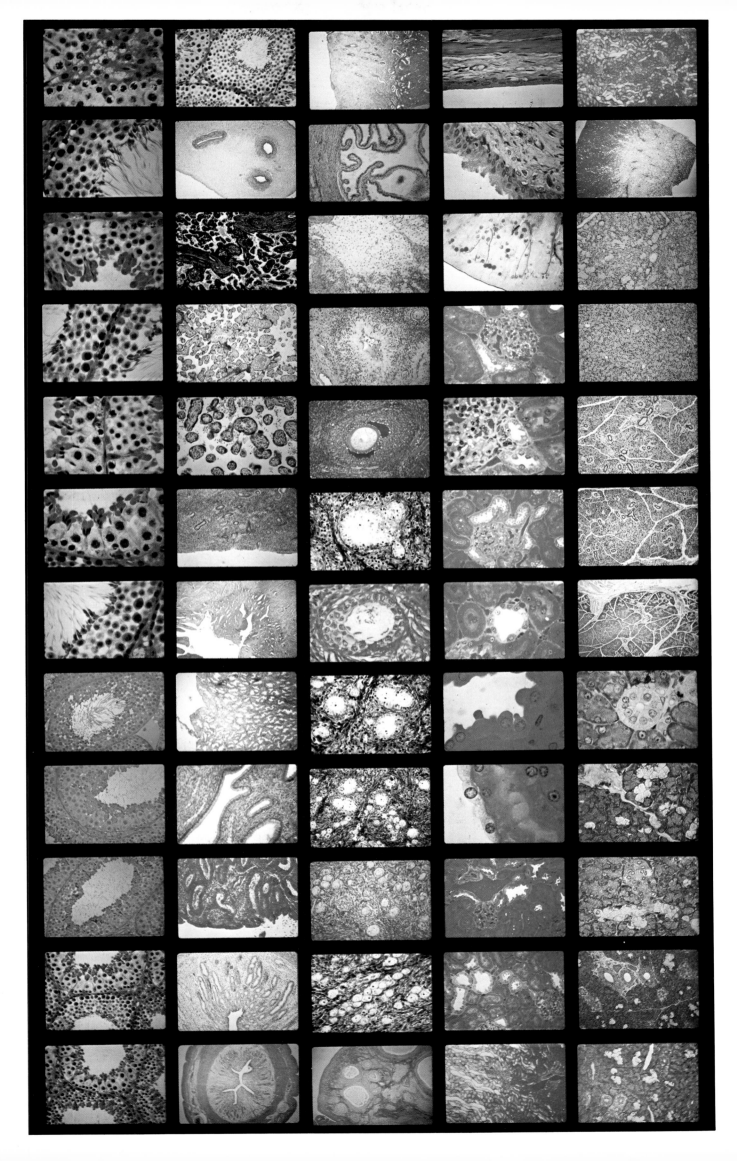

Interior Landscapes: Pavel Tchelitchew

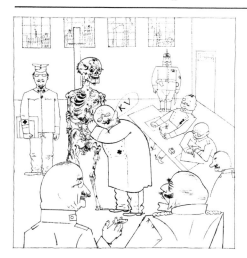

Pavel Tchelitchew (1898-1957) referred to his anatomical paintings as his "interior landscapes" of man. This series of paintings, which includes different sections of the body as well as whole figures, shows the range of his anatomical knowledge as well as his skill as a painter. Executed about 1943, these paintings are among Tchelitchew's later work and carry the weight of his many years as an artist pondering the precarious nature of man's existence.

Tchelitchew was born in Moscow to a family of Russian aristocrats. His early interest in the human body expressed itself through his desire to dance ballet. His father, however, forbade him to become a dancer, and so he turned to his second choice, which was to be an artist. The chaos of the revolution in Russia in 1917, combined with his need to express his ideas freely, drove him from Russia, first to live in Berlin and then, years later, to Paris. He remained in Paris until he moved to the United States in the 1930s.

Tchelitchew's work took many different directions. He did stage design in Russia, continuing to work in Berlin and Paris in the same area. He eventually came to work with the brilliant Serge Diaghilev. It wasn't until the late 1920s that he began to take his painting seriously.

His preoccupation, as an artist, with human anatomy has often been compared to that of Dürer. Theirs was an artistic tradition rooted in years of schooled observation of the human figure. Certainly Tchelitchew's painting and Dürer's work were as unlike as any two styles could be. It is their shared focus on and intimate knowledge of human anatomy that underlie the comparison.

Tchelitchew's "interior landscapes" of the human form are as precise as they are complex. His skill at superimposing body system upon body system is remarkable. He articulated each individual detail of each system despite all the layers. It is as if each bodily system represented a whole subject unto itself.

Tchelitchew was clearly fascinated by man's structure, but not purely as an intellectual concept. His figures have a luminous inner glow, a curious sense of warmth that light up their "landscape." These anatomical figures say something about the shape of man's body, and even more important, they say something about Tchelitchew's feelings about the character of man's essential nature.

OPPOSITE: *Anatomical Painting*, c. 1945, by Pavel Tchelitchew. Collection Whitney Museum of American Art, New York. ABOVE: *Fit for Active Service*, 1916, by George Grosz. Collection The Museum of Modern Art, New York. A. Conger Goodyear Fund.

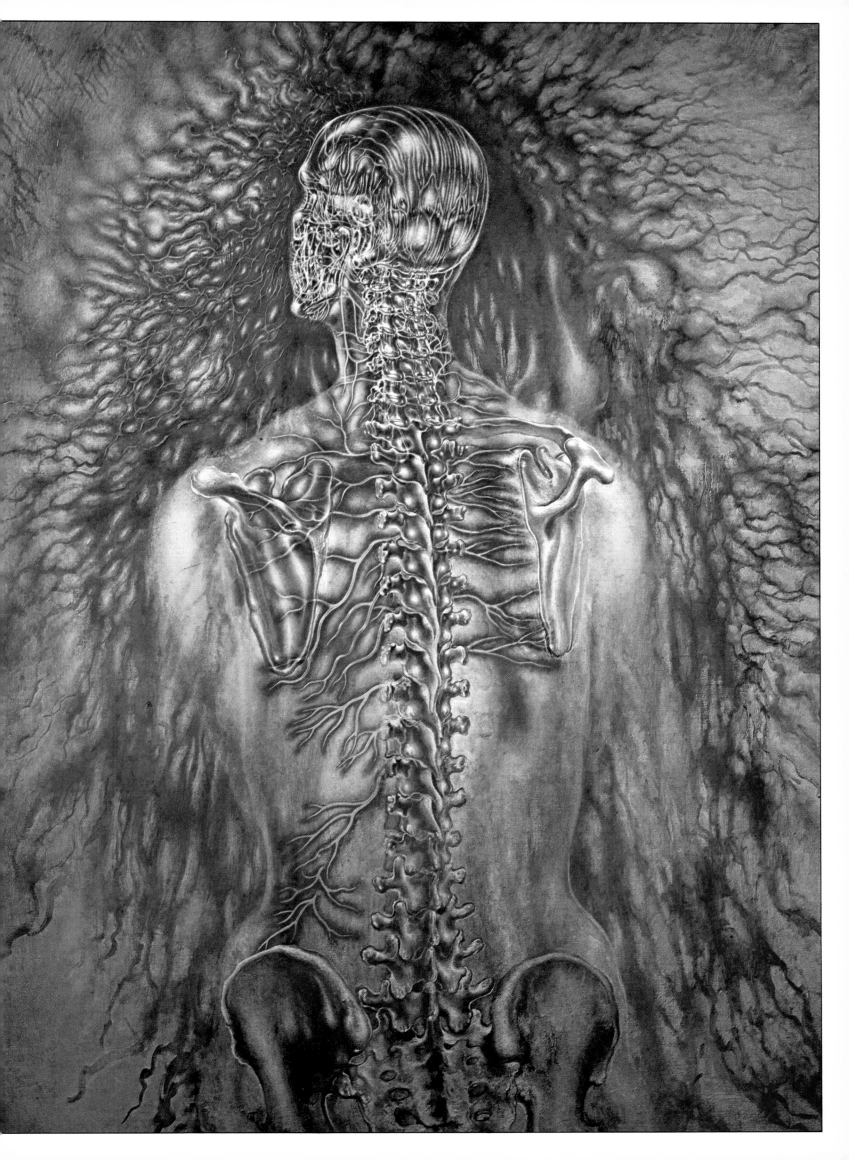

Photographic Visions: Lennart Nilsson

Among the twentieth century's most visionary works in human anatomy has to be that of the Swedish photographer Lennart Nilsson (1922-   ). Nilsson's photographs of the interior landscape of the human body reveal images that are as mysterious as they are beautiful. He has entered regions of the body never seen by the naked eye. Lennart Nilsson's photographs create a startlingly complex portrait of the secret world beneath the surface of man.

Using the most modern medical techniques and optical systems available, Nilsson has explored every crevice of the human body. His pictures of the heart and the blood vessels are the very first of their kind. His images of the reproductive system and the fetus *in utero* astounded scientist and layman alike when they were first published in the 1960s. To capture some of these images, Nilsson used a light microscope to magnify tissue only a fraction of a millimeter thick up to one thousand two hundred times. He has worked with electron scanning microscopes whose depth definition is five hundred times that of the light microscope. When none of the techniques available seemed to accommodate his needs, Nilsson invented his own systems.

Nilsson's photographs not only show the structures of the human body; they also help to explain how the complex machine of man works. In his series of pictures on the reproductive system, he begins by showing the tiny spermatozoa meeting and fertilizing the egg. Next come the photos of the first cell divisions, the embryo in different stages of development, and, finally, the awesome image of the fetus attached to the placenta at about seven months. He has literally followed the step-by-step creation of another human being. Nilsson has applied the same attention to many other systems in the body.

To the beholder of Lennart Nilsson's work, description is superfluous. These photographs are powerful statements in their own right. The structure of the human body is strong material, but it demands someone with Lennart Nilsson's skill and inspiration to tackle its complexity and mysterious beauty.

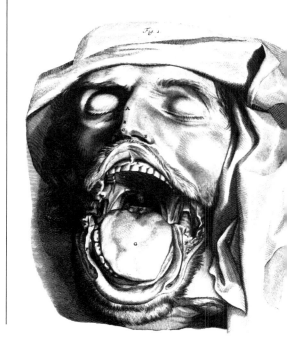

OPPOSITE: A section through a tooth as seen through an interference microscope, 1973, by Lennart Nilsson. ABOVE: Another view of the teeth, probably etched from a book on surgical procedures, eighteenth century, Manchester Infirmary. Yale Medical Library.

Thermographic Imaging

Modern science in the twentieth century has developed many new techniques for producing images of the interior structures of the human body. These imaging techniques have revealed a vast amount of new information about the way the human body works as well as details about its anatomy. Dr. William Harvey, in the seventeenth century, was the first anatomist to consider the relationship between the structure of the body and the way the body functions. Since then, with the proliferation of new imaging techniques, including the invention of the microscope, there has been a shift of focus in the study of human anatomy from the isolated observation of structure to the more complex issue of the functions of its structure.

One of the more recent imaging techniques is thermography. More like a painting than a medical image, a thermograph is a picture of the heat structure of the body. An image of the heat structure is produced when an electronic camera scans the body from side to side, reading its infrared radiation. Infrared is an area of heat radiation on the electromagnetic spectrum which is not visible to the naked eye. The eye is tuned to only a narrow band of the electromagnetic spectrum, that whole range of radiant energy which all living things give off. Radiation, the process whereby energy is emitted as light-waves or particles, can be measured by several of the new imaging techniques. Thermography measures the heat radiation of the body and transforms it into a signal somewhat like the one used for television transmission. The signal is fed into a computer which codes the temperature into colors and projects the image onto a video screen. The colors of green and blue represent the coolest sites; the colors of white and yellow signify the hottest areas. The final image represents man's and the computer's combined capability at creating a system that translates what is invisible to the human eye into a visible language.

The body's heat structure can describe its condition in several ways. For example, the presence of a cancer may be indicated by the fact that one part of the body gives a different temperature reading from the area surrounding it. Among other things, thermographs can detect vascular problems, the depth of burns and the resultant tissue damage. It is also possible to observe the way the body loses and retains heat.

The field of human anatomy has benefited greatly from all the new types of body images. These images provide a few more clues which may help unravel the mystery of the human body.

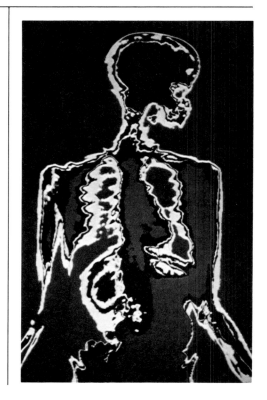

OPPOSITE: A thermograph of a four-year-old girl showing the heat distribution in her body, 1970s. ABOVE: Another new technique for examining the interior structures of the body is the density scan, late 1970s. Woodfin Camp/Howard Sochurek photos.

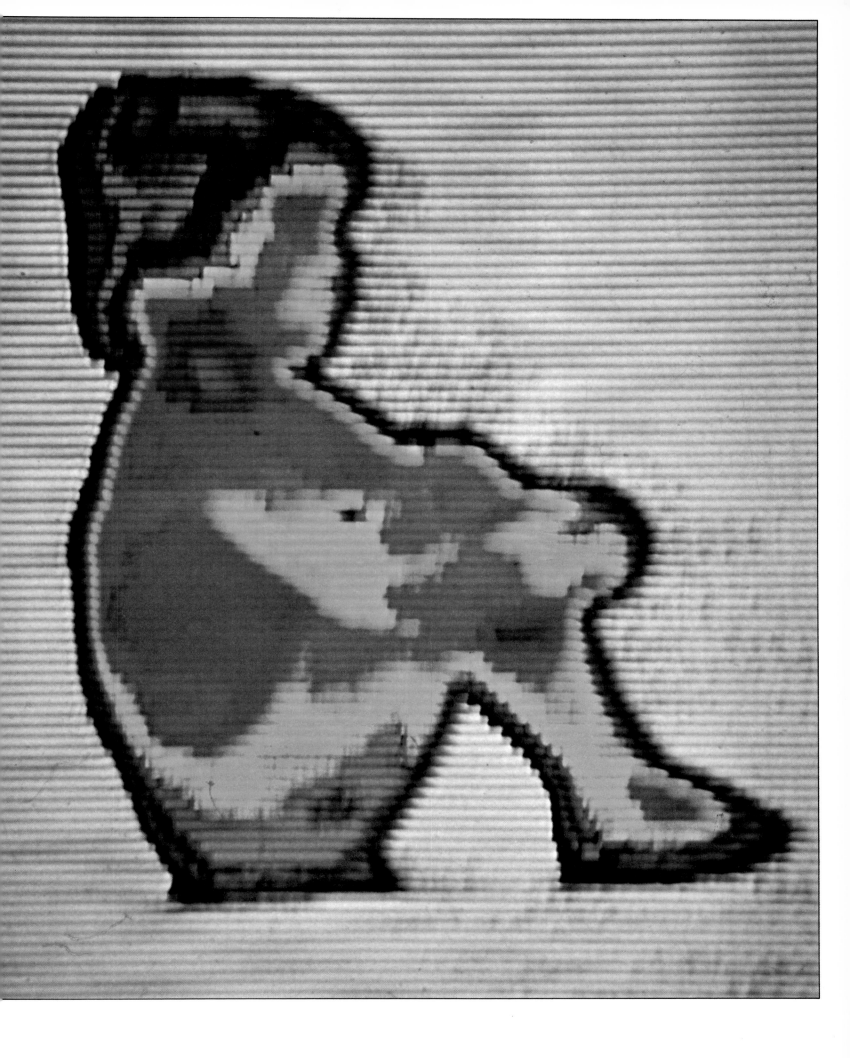

Gamma-ray Imaging

With the technological advances in medical imaging it is now possible to examine the structure of the human body for early warning signs of danger. Whether it be a warning of cancer, an early diagnosis of lung disease or an early signal of bone disease, the computer has enabled medical technology to probe the body's innermost space for diagnostic purposes. It is not difficult to appreciate what advances in this type of body imaging mean to the quality of human life in the prevention of sickness and disease.

One of the most recent techniques for imaging human structure involves reading the invisible gamma-rays emitted by isotopes that have been introduced into the body. Isotopes are chemical elements in a form that gives off high-frequency, penetrating radiation known as gamma-rays. An isotope is either ingested or

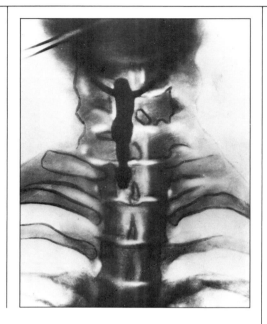

injected into certain areas of the body. It then collects in these specific areas, and an electronic eye scans the body and locates it. The scanner transmits the gamma-rays the isotope emits to a computer which translates the message into a color-coded image. The density of gamma-rays outlines the normal or abnormal dimension of the area in which the isotope has collected. The computer projects an image of the radioactive material that can be studied for any symptoms or abnormalities.

Before gamma-ray imaging was available, medical science did not know the potential of such a precise and painless diagnostic procedure. Either a problem went undiagnosed until it was at a much more critical stage, or there was the possibility of painful exploratory surgery. As man's inner structure has become more accessible through new techniques of imaging, the chances of better health have improved considerably.

OPPOSITE: This gamma-ray image shows the active bone growth in a scan of the upper torso, 1970s. Photo courtesy of Elscint (GB) Ltd., Great Britain.
ABOVE: An entirely different view of human bone structure is this early x-ray photo revealing a crucifix lodged in a woman's throat, 1924. United Press International Photo.

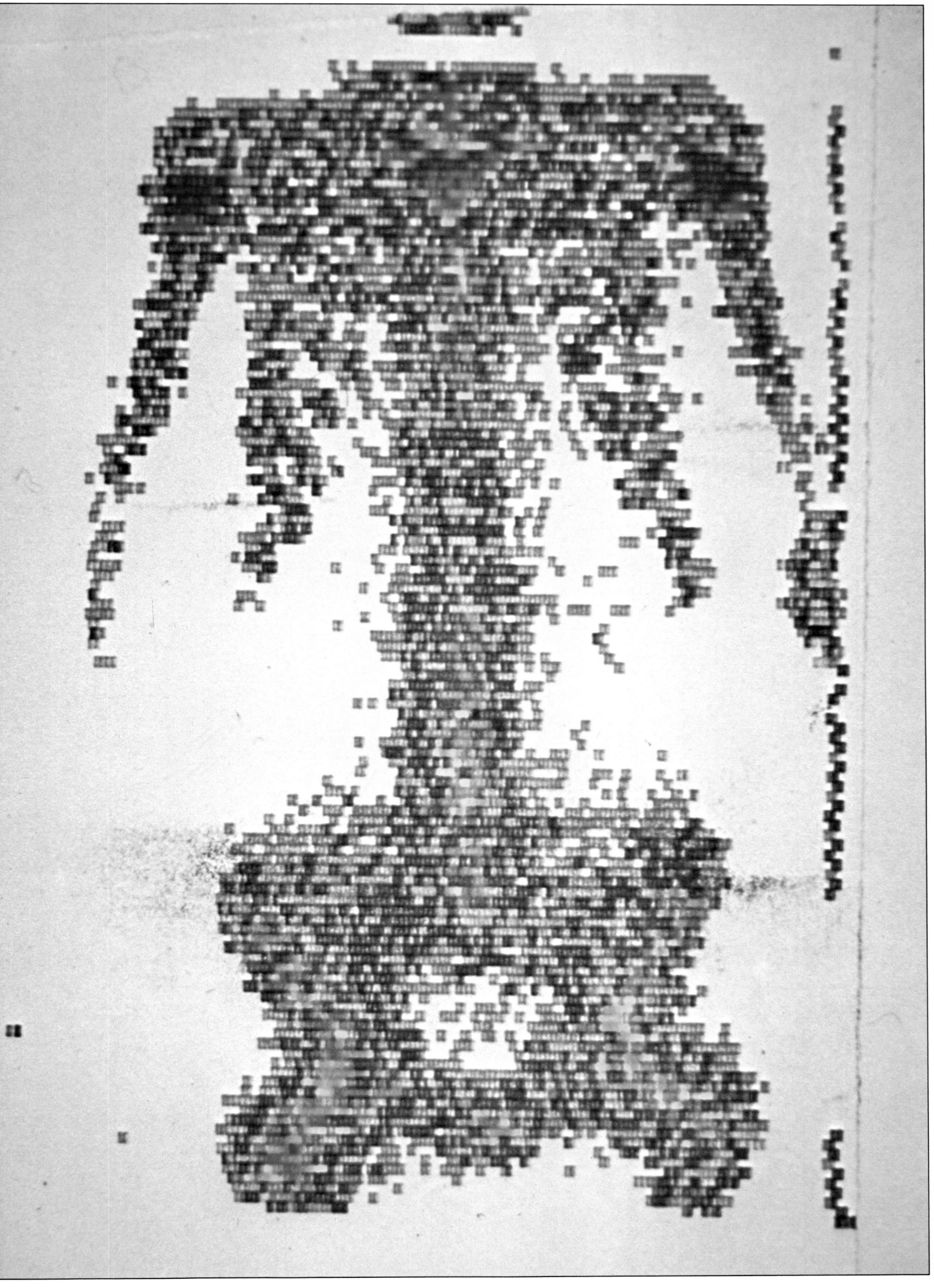

Ultrasonography

In addition to the various other twentieth-century advances in medical imaging, there is now a procedure called ultrasonography which can produce an image of a live human fetus at varying stages of development inside the womb. The theory behind ultrasonography originated with the sonar systems in submarines. Just as the submarine's sonar probes the ocean for the dimensions of objects in its path, so ultrasound probes the body's tissue for the dimensions of objects in its range. Recently, there have been new developments in the usage of ultrasound for diagnosing cancers and disease in other areas of the body.

Ultrasonography works on a simple and safe principle. A transducer, which is a probe at the end of a long arm, is laid against the individual's body. The transducer emits ultrahigh-frequency sound waves at over one thousand cycles a second. These sound waves pass through the body, are reflected off whatever structure they are aimed at and resound as a series of echoes. These echoes are then picked up by the transducer and relayed to a computer which reassembles them into an image. The echoes are so accurate that doctors can, for example, distinguish soft tissue from muscle. As far as anyone knows, these ultrahigh-frequency sound waves do not disturb or harm the body in any way.

With the aid of ultrasound, doctors can now follow the progress of a human fetus from shortly after conception to the time of its birth. This means that for the first time, any symptoms of trouble that might show up in the fetus can be identified and treated directly instead of through the mother. The echo image enables doctors to observe the external features as well as the movements of the fetus. They can monitor its growth by means of special measurements designed to fit this type of image.

For the first time in history, the world can observe the anatomy of a human fetus as it grows within the safe, ocean home of the womb.

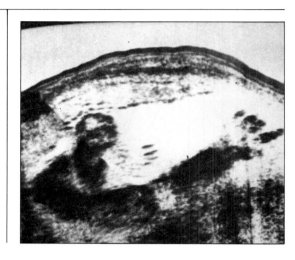

OPPOSITE: An ultrasound image of the fetus entering the birth canal, 1978. Courtesy of John C. Hobbins, M.D. (The Williams & Wilkins Co., Baltimore). LEFT: An earlier depiction of the fetus in the womb, 1626, *De Formato Foetu Liber*, by Adriaan van den Spieghel. Yale Medical Library. ABOVE: Another ultrasound image of the fetus, about full-term, 1978. Courtesy of John C. Hobbins, M.D. (The Williams & Wilkins Co., Baltimore).

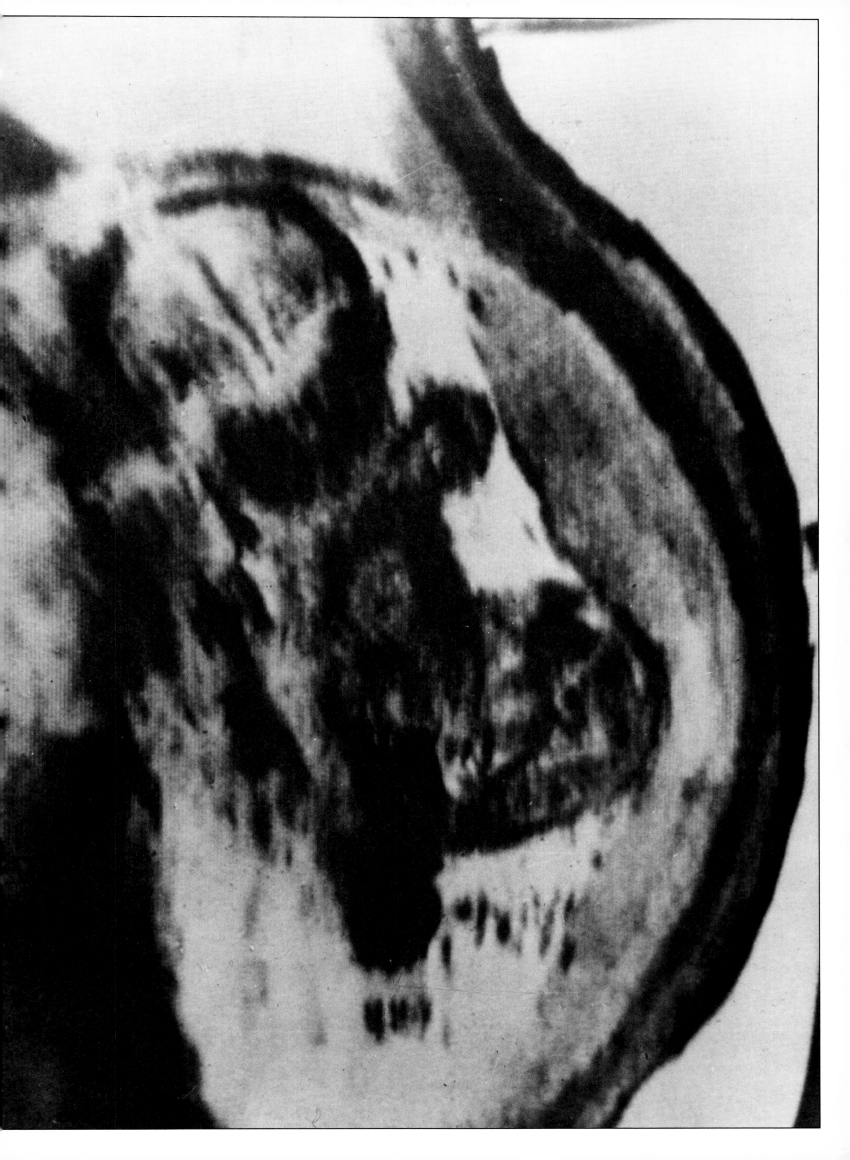

Anatomical Interpretations: Leonard Baskin

"Our human frame, our gutted mansion, our enveloping sack of beef and ash is yet a glory."   Leonard Baskin, 1959

Leonard Baskin's (1922-   ) anatomical studies of the human form are a departure from the artistic anatomies of previous centuries. His portfolio of drawings titled *Ars Anatomica* is one of the few serious anatomical works by a well-known artist in the twentieth century. Yet, unlike the classical artistic anatomies of the past, his drawings do not represent a thorough examination of the body's structure, nor do they make any pretense of scientific accuracy. Far from that, Leonard Baskin's anatomical studies are his private observations of man in a physical form.

Leonard Baskin deeply believes in man's exploration of man. The physicality of the human form intrigues him, inspires him, mystifies him. His portrayal of that physicality comes out of his investigations of man and expresses Baskin's concern about the quality of his existence. He has said, "Man must rediscover man, harried and brutalized, distended and eviscerated, but noble withal, rich in intention, puissant in creative spur, and enduring in the posture of love."

As Baskin interprets the human form in his drawings, there is a heartfelt respect for man that runs through all his work. He will draw a vacant, eyeless skull and its existence becomes a reminder of the fullness and warmth in the living eyes of man. He renders the chest split open and stretched back to reveal the cavity beneath the surface. Instead of focusing on grotesque images of death, he somehow evokes the contrast of a chest throbbing with the breath of life. Leonard Baskin dissects the human structure, but only as it serves his purpose of illuminating life by holding it up against the dark images of death.

Baskin's woodcuts and lithographs represent a unique and vital interpretation of the human form. In the twentieth century already plied with mysterious, abstract medical images, Baskin's anatomical drawings are reassuring in their humanity and their concern for man.

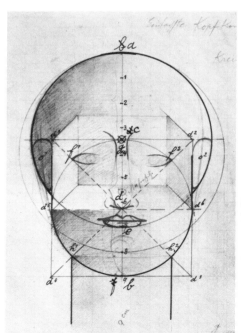 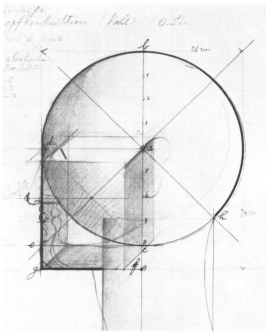

OPPOSITE: *The Anatomist* (n.d., c. mid-1900s) by Leonard Baskin. Collection The Museum of Modern Art, New York, Gift of the Junior Council. LEFT: Two studies of the head done by Bauhaus scholar Oskar Schlemmer, c. 1928. Courtesy of Tut Schlemmer.

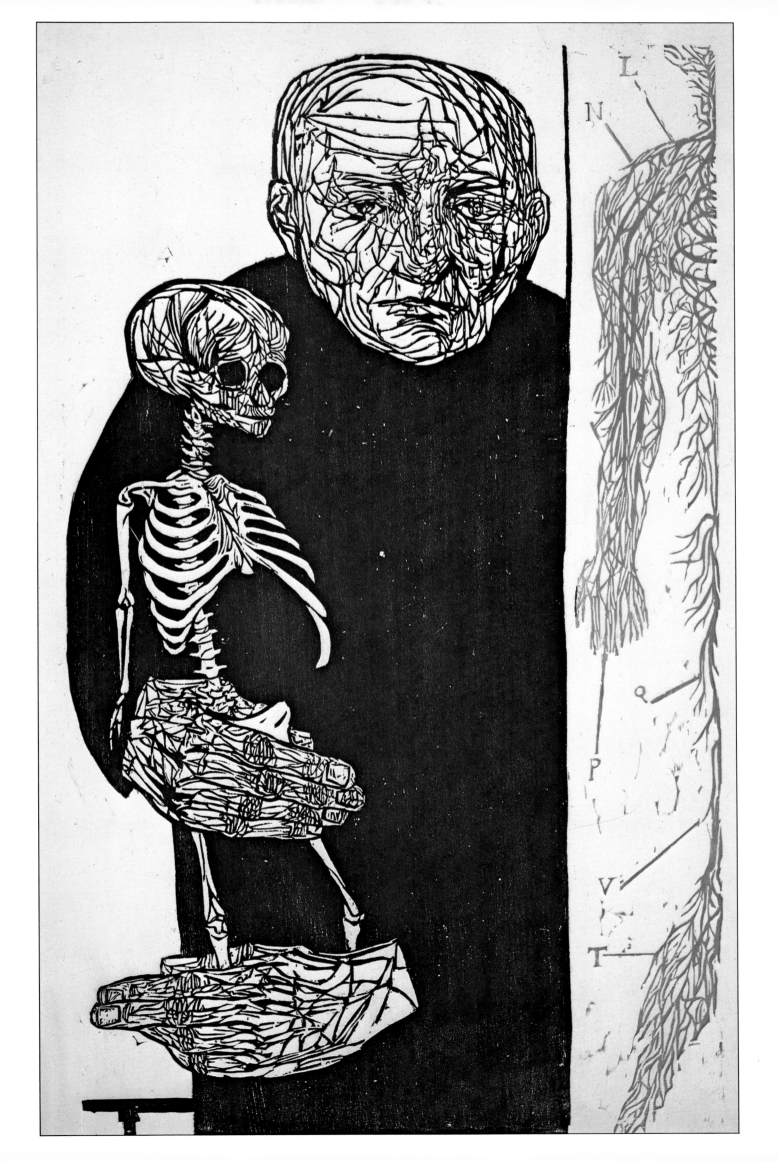

X-ray Imaging

The structure of the human body has probably become more familiar to the modern world through the widespread use of x-rays than it has through any other single imaging technique. By and large, x-rays are still the technique most frequently relied upon for diagnosis and treatment of various ills of the human body. Because they can penetrate the walls of the body, x-rays illuminate the details of the body's interior structures. X-ray images have proved highly valuable to medical science in its continuing efforts to explore and understand the human body.

X-rays were discovered in 1895 by Wihelm Konrad Roentgen. X-rays travel, as light does, in waves. Their wavelength is, however, shorter than that of light and can penetrate solid substances in a way that light cannot. An x-ray differs from light in that it is an area of radiation on the electromagnetic spectrum that is invisible to the naked eye.

The technique of producing an x-ray image of the body begins when a stream of electrons is shot through the body at a high velocity, causing the electrons in the body to react and emit x-rays. These x-rays then act on a special film to produce an image. The power of x-rays depends on two factors. One is the density of the substance through which the x-ray must pass. The other is the speed at which the electrons entered the body.

X-rays affect film in much the same way in which light affects it. Where the body area is not so dense, in the soft tissues, for example, more x-rays pass through it to the film and the film appears black. Where the body area is more dense, as in the case of the bones, fewer x-rays reach the film and the film appears white. The final image reflects the human struc-ture with precision and detail enough for the doctor to make a diagnosis of the body's condition from the film.

X-rays are powerful, and they can cause severe damage to the body if it is exposed to too many of them. Over the past few years, medical science has discovered that heavy doses of x-rays which can kill healthy tissue can also be used to destroy cancerous tissue. X-rays are now being used, on this principle, to treat a number of cancers.

The most recent advance in the use of x-rays has been the invention of tomography. Hailed as the biggest single advance in diagnostic medicine since the discovery of x-rays themselves, tomography is a method of using x-rays to photograph one specific plane of the body. In this procedure, the body is placed in a frame. Within the frame, there is an x-ray tube mounted across from a series of sensitive crystal detectors. A beam of x-rays fans itself through the body at a series of angles about 10 degrees apart. The detectors pick up the x-rays and feed the information into a computer which assembles it and transforms it into an image. This image can then be analyzed, and within a short time a doctor can detect an irregularity almost anywhere in the body, from the brain to the waist and on down.

X-rays, along with many of the other recently developed imaging techniques, have had a great impact upon the study of human anatomy. These imaging techniques have opened up new frontiers within the body. No longer confined to the isolated observation of structure, the study of human anatomy has been forced to widen its scope to include a look at how the body functions. The science of human anatomy has remained vital just because it has taken up the challenge and dared to explore these new vistas of man.

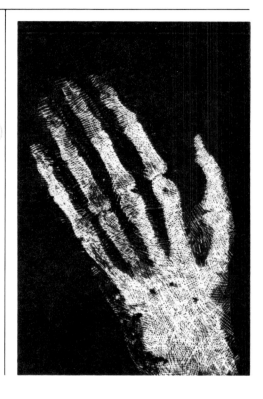

OPPOSITE: A color-coded x-ray of a healthy woman wearing a necklace, 1970s. Photo courtesy of Agfa-Gevaert, West Germany. The color in this image is created by means of a photographic technique using Agfacontour film. ABOVE: This hand loaded with shotgun pellets was x-rayed in 1898 by a Thomas Edison machine prior to an operation to remove the bullets. Courtesy of the General Electric Corporation.

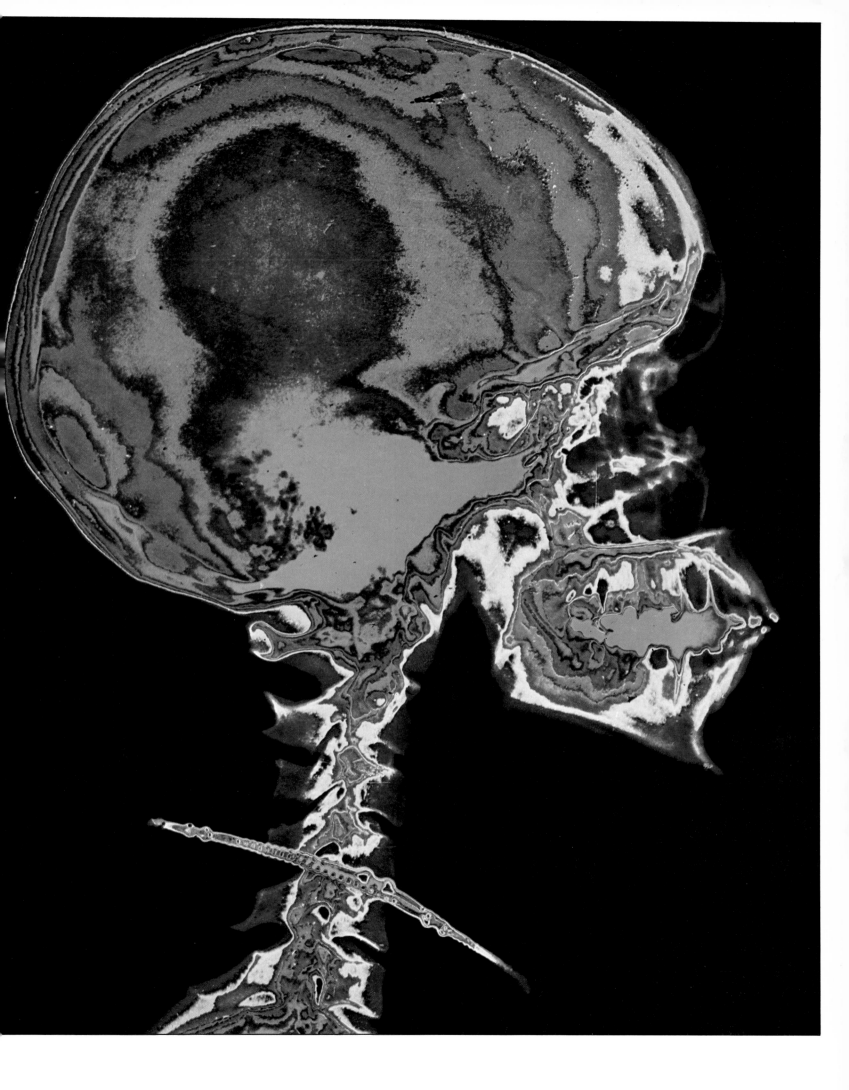

Acknowledgments

We would especially like to thank Frank
Gyorgyey and William Guth of Yale University
for their invaluable suggestions and help in
gathering material for this book. We are also
grateful to Gordon Mestler for his kindness
in permitting us to use material from his
collection. His enthusiasm for this project
was an inspiration. Finally, we would like to
mention the assistance provided by Mrs.
Alice D. Weaver and her staff in the Rare
Book Room of The New York Academy
of Medicine.

We wish to acknowledge the assistance of
the New York Public Library for the material
on pages 10, 34 and 53.

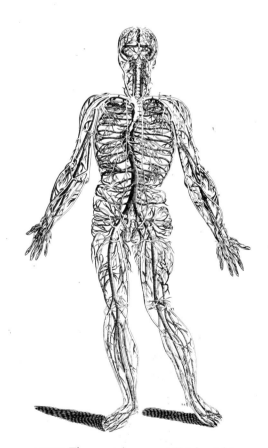

ABOVE: The arterial system, 1626, by Adriaan
van den Spieghel. Yale Medical Library.